S0-CVG-250

Democracy and Economic Change in India

C H I N A
CEASE-FIRE LINE
KASHMIR
AFGHANISTAN
PAKISTAN (WEST)
Indus R.
PUNJAB
Delhi
SIKKIM
NEPAL
BHUTAN
Brahmaputra R.
UTTAR PRADESH
Ganges R.
ASSAM
RAJASTHAN
I N D I A
BIHAR
PAKISTAN (EAST)
BURMA
WEST BENGAL
Calcutta
GUJERAT
Ahmedabad
MADHYA PRADESH
ORISSA
MAHARASHTRA
Bombay
Poona
BAY OF BENGAL
ARABIAN SEA
Hyderabad
ANDHRA PRADESH
GOA
MYSORE
Bangalore
Madras
Coimbatore
MADRAS
KERALA
Madurai
CEYLON
0
200
400
Statute miles

INDIA, 1964

Democracy and Economic Change in India

by

George Rosen

University of California Press
Berkeley and Los Angeles

1967

University of California Press
Berkeley and Los Angeles, California

Cambridge University Press
London, England

New Edition, with Corrections and Additions, 1967
Library of Congress Catalog Card Number: 66-13986

Printed in the United States of America

To
Kusum and Mark
and
Sachin Chaudhuri

"We've missed you. As I say, people have gone serious lately, while you've just been loafing about the tropics. Alastair found something about Azania in the papers once. I forget what. Some revolution [there]. . . .

"Can't think what you see in revolutions . . . [it] doesn't make much sense to a stay-at-home like me. . . .

"Write a book about it sweety. Then we can buy it and leave it about where you'll see and then you'll think we know."

Evelyn Waugh, *Black Mischief*

I have discovered that my preference as a reader is for the unitary, though partial, vision of a grand theme by a single pair of eyes, rather than for a series of minor, though definitive, variations on it. It is in this spirit that I have approached the writing of this book.

A. Shonfield, *Modern Capitalism*

If we must talk of social evolution, we ought to remember that it takes place through the action of human beings, that such action is constantly violent, or merely short-sighted, or deliberately selfish, and that a form of social organization which appears to us now to be inevitable, once hung in the balance as one of several competing possibilities.

R. H. Tawney, *The Agrarian Problems in the Sixteenth Century*

Preface

This study arises directly out of my economic work, which has now extended over more than ten years in and about India and Nepal. Three of those years were spent in India. The last two years of my actual stay in that area were as an economic advisor to the Government of Nepal and to the Government of the State of West Bengal. In that period I was struck by the relationship between politics and economic development, and I saw that the easiest part of the process of development is to prepare a "Plan" for economic development. More difficult is to create the atmosphere in which the Plan will be accepted, and even more to begin to take the steps to implement it. Any Plan is part of the politics of the country, and it will be accepted or implemented only within a political environment that accepts the desirability of economic development and specific economic policies to achieve it. If such an environment exists, a country is a long way toward economic development. This study of the relationship in India between the political environment and economic development is my personal response to my experiences as an observer of the Indian scene.

The serious question arises whether enough is known of what is happening in India concerning the relationship between political and economic change. From my stay there I am convinced that almost no generalization is possible for the entire country and that it is possible to find several exceptions to any generalization that might be made. I should not have attempted this study ten or even five years ago. But since 1955 there has been a whole series of more or less isolated village studies by anthropologists in particular regions of India; and, somewhat surprisingly, from these village studies it is possible to discern a crude pattern of both the social structure of Indian rural life and of changes in that structure since 1947. Also, various anthropologists have developed theories of Indian social structure that are meant to apply on a national basis. If I were a professional anthropologist I doubt that I would be so foolhardy or courageous as to apply those individual village or area studies to national generalizations, but as an economist experienced in dealing with macroeconomic problems I have been willing to take the risk; and I hope my anthropologist friends will not find it too ludicrous.

Similarly, on the side of political science and economics, research has been under way in various areas, and a flood of statistics—good, bad, and indifferent in quality—has been forthcoming since 1950. Many of these data have not been organized in a fashion to answer my questions; many of them do not extend over a long period of time. But there are now enough to reach some conclusions with respect to trends, even if the exact results are indefinite. This study hopefully points out gaps in statistics and indicates the types of additional data that must be collected if the questions I have raised are to be answered.

Concerning the acceptability of the results I can best refer to some remarks by Keynes on a paper by A. K. Cairncross, as noted by the latter with his reaction in the Preface to his book, *Home and Foreign Investment, 1870–1913*.

> Keynes prefaced his comments on [the paper] by a reference to the origin of the Septuagint. He thought that the kind of history which I had presented to the Club should be submitted to a similar test before being accepted as holy writ. If seventy statisticians could be locked up, like the Hebrew scholars, in separate cells and each emerge with the same statistical series, it would be possible to accept their results without further question. Failing such proofs of inspiration and truth, one could only accept the plausible and reject the implausible. This was a pronouncement from which I did not and do not dissent. The reader should look on what follows as one step exactitude and reflect on the sixty-nine steps that are most unlikely ever to be taken [by the author himself].

I joined The RAND Corporation in December, 1962, after a stay in India, and at RAND I have become much more aware than previously of the implications of the questions I have raised for U.S. policy both with respect to India itself and for other underdeveloped countries. I have therefore put this study in a broader framework than just India—first pointing out the relevance of the relationship between political change and economic change in underdeveloped countries to the United States, and then presenting in Appendix A a general framework within which I believe this relationship can be explored. The body of the book then applies this framework to past Indian development and the prospects for the future, and it draws some implications for U.S. policy toward India on the basis of previous analysis. In a brief concluding Epilogue I have indicated some of the possibilities of applying this analysis to problems of other underdeveloped countries than India.

Acknowledgments

Since this book does reflect an absorption in India that goes back many years, it is especially difficult to distinguish the many debts of friendship and assistance. So much is part of my experience and has been so deeply absorbed that it is impossible to separate into its components. My initial interest in India goes back to the late Frank D. Graham of Princeton who stimulated and guided my doctoral thesis on the industrialization of the Far East. My first two years in India were with the Massachusetts Institute of Technology's Center for International Study. Max Millikan, the director of the Center, Wilfred Malenbaum and Paul Rosenstein-Rodan, the directors of the India Project, and Sukhamoy Chakravarty, Richard Eckaus, Helen Lamb, Louis Lefeber, and Walter C. Neale, all of whom have at one time or another been associated with the Center's Project in India or Cambridge, have been most helpful as stimulators and critics of my ideas; Myron Weiner's willingness to exchange ideas and works in progress on the political changes in India has provided both a flood of new ideas and a constant check to my own ideas in this area; Harold Isaacs, Lucian Pye, and Jean Clark, although not specifically associated with the Center's India project, have provided continued support and insights to my work.

My two years of residence as advisor in India and Nepal were for the Ford Foundation. Douglas Ensminger, the representative of the Foundation in India, was the friend and supporter who made the experience possible; Morton Grossman, the Ford Foundation economist, has a unique knowledge of Indian economics; my colleagues of the Ford Advisory Groups in both Nepal and Calcutta provided interdisciplinary stimuli, experience, and argument without which the subsequent study would have been impossible; Wolf Ladejinsky, an old friend formerly with the Ford Foundation, has always been a source of experience and insights; David Hopper, the economist associated with the Ford agricultural program, provided stimulating hours of his ideas on Indian agriculture.

To list the Indian friends whom I got to know well in Bombay, Calcutta, and New Delhi during my four years' stay would be impossible. They have included people in all walks of Indian life; without their kindness and friendship, their hospitality and conversation about their country and its hopes, this study would have been impossible. My conclusions are my own; but the ideas have grown, been changed, been honed and refined by exchanges over four years. Among these friends in Bombay, and apart from any of those mentioned elsewhere in these pages, I would like to mention A. D. Gorwala, Mr. and Mrs. S. Natarajan, Mr. and Mrs. Daniel Thorner, and Mr. and Mrs. Evelyn Wood. My dedication of this book to Sachin Chaudhuri, one of my closest friends, and one of India's genuinely great men, at whose home and in the pages of whose magazine, *The Economic Weekly,* I have met so many of these friends, can only be a small token of my affection and respect for them. I hope that they do not find this book the work of an "amateur"; nor that they find it an ungrateful response to the hospitality I have received.

The opportunity to write this book I owe to The RAND Corporation as part of its research program for the Office of the Assistant Secretary of Defense, International Security Affairs. I should like to thank both of these organizations. Within RAND, Burton Klein, the head of the Economics Department, accepted the idea and supported it over two years and made possible three short trips to India since 1962. Drafts of the book have been criticized in whole and in detail by Ian Graham of the Social Science Department, Edmund Dews, Evsey Domar (a 1964 summer visitor), William A. Johnson, Richard R. Nelson, James R. Schlesinger (especially Chapter 13), Helen B. Turin, and Charles Wolf, Jr. of the Economics Department. Apart from their detailed criticisms of the draft they provided encouragement and a continual sounding board for the changing ideas. This has been essential for the development of this study. Others of my colleagues have criticized papers or chapters that I have completed. I should especially like to thank Alan Carlin, Joseph M. Carrier, Jr., Luigi Einaudi, Allen Ferguson, Brownlee Haydon, John Hogan, Benton Massell, Horst Mendershausen, Mancur Olsen, and David Wilson for their criticism and insights.

Apart from my colleagues at RAND, various friends have been kind enough to encourage this study, to read it, and to criticize earlier outlines and chapters. I should especially like to thank Fred G. Bailey, University of Exeter (England); Joseph Berliner, Brandeis University; Flournoy Coles, System Development Corporation; Hampton Davey, University of California at Los Angeles; Scarlett Epstein, Royal College of Advanced Technology, Salford; John Kenneth Galbraith, Harvard University; John Hitchcock, University of California at Los Angeles; Richard Lambert, University of Pennsylvania; John P. Lewis, Agency for International Development, New Delhi; Charles E. Lindblom, Yale University; Michael Mazer, Harvard University; S. Nath, Hindustan Lever; Baldev Raj Nayar; Hans Ries, Continental Ore Corporation; Evelyn Ripps, Agency for International Development, Washington, D.C.; Henry Schloss, University of Southern

California; M. N. Srinivas, University of Delhi; Prakash Tandon, Hindustan Lever; Howard Wriggins, U.S. State Department; Peter Wright, International Bank for Reconstruction and Development; and Maurice Zinkin, Lever Brothers (England). I have also had the opportunity, through John Hitchcock's support, to present some of these ideas as they pertain to India to a graduate class in Anthropology at the University of California at Los Angeles for one term. This discipline contributed to shaping the work as it went along.

Finally, this would not have been written except for Doris Carlson, who typed and supervised the typing, deciphered my handwriting, corrected my grammar, and met my deadlines. Dee Yinger and Jean Arensburger checked references and footnotes and eliminated many errors.

As a final word, this manuscript was largely completed by the end of 1964. Since then several major events have taken place—the fighting with Pakistan over the Rann of Kutch and Kashmir in mid-1965, the unusually poor crop of 1965 following a good crop of the previous year, and the death of Prime Minister Shastri and his succession by Indira Gandhi. I have made a minimum number of changes in the page proofs to take these events into account. But this book was written as a study of more underlying trends in the Indian society and economy, and I strongly believe that the trends described, and the problems stressed, have not been changed.

Needless to say, I alone am responsible for the interpretations and conclusions and for any errors that may remain.

George Rosen

Contents

PART IV: ECONOMIC POLICY AND ACHIEVEMENT, 1947–1962

List of Tables

Appendix

PART I

Introduction

The most fruitful areas for the growth of the sciences were those which had been neglected as a no-man's land between the various established fields. . . . It is these boundary regions of science which offer the richest opportunities for the qualified investigator.

N. Wiener, *Cybernetics*

It should be one of the main tasks of applied economics to examine and unravel the complex interplay of interests, as they sometimes converge, sometimes conflict. . . . In the first place, we should want to know where interests converge, for in these cases we could make at once generally valid recommendations. We should also want to ascertain where lines of interest intersect. In these cases we could offer alternative solutions, each one corresponding to some special interest. Both types of solutions can claim objectivity.

G. Myrdal, *The Political Element in the Development of Economic Theory*

1
The Theme

This is a book about the political revolution that occurred in India in 1947, and the associated political, social, and economic changes that have followed from the upheavals of that year. The defining characteristic of a revolution is not violence or its lack, but rather the speed of change and the degree to which it occurs throughout the entire society. Within the short period of fifteen years, India has undergone a series of changes that have spread through the entire society although they were initially confined to the political sphere. In the process the changes in the political sphere have been interdependent with the changes in the economic and social spheres, and the past interaction has been such that it influences the possibility of future changes in all spheres.

I shall lay greatest stress upon the economic changes; economic policies and programs provide the focus of this study. But the basic problem I deal with is that of the relationship between the political and economic changes. The relationship between economics and politics is closest in the policy field. I shall therefore specifically look at how the political changes of Indian Independence led to shifts in political and economic power and thus to the economic development policies of the Plans; how these policies in turn influenced the distribution of political power since Independence; and finally how the changing character of political and economic power influences potential economic policies for the future of India.[1]

[1] Until recently, modern economics did not treat the political and social changes associated with economic policy or change, and confined itself more or less to analysis of market behavior or macroeconomics within carefully limiting assumptions with respect to economic and social change. As economists have become involved in the problem of economic development, they have inevitably looked into non-economic factors associated with development. Recent examples of such works by economists are: Everett E. Hagen, *On the Theory of Social Change* (Homewood, Illinois: The Dorsey Press, Inc., 1962); Albert O. Hirschman, *Journeys Toward Progress* (New York: Twentieth Century Fund, 1963); Raymond Vernon, *The Dilemma of Mexico's Development* (Cambridge, Massachusetts: Harvard University Press, 1963); Paul A. Baran, *Political Economy of Growth* (New York: Monthly Review Press, 1957). Walt Rostow, A. Gerschenkron, and J. Habbakuk are contributing significantly to the understanding of development, and the past work of R. H. Tawney and Max Weber on early European economic history is now seen as relevant in analyzing growth problems in non-European countries. For a useful analysis of the trend of interest of economists in these political and social problems, see Joseph Dorfman *et al., Institutional Economics: Veblen Commons, and Mitchell Recon sidered* (Berkeley: University of California Press, 1963), especially the lectures by R. A. Gordon and J. Dorfman.

This study treats India since I know that country best; but the general framework of analysis will, I hope, be useful for analyzing the problems of the political economy of growth in many other countries as well, and it offers possibilities for the comparative analysis of these problems.

But apart from its value for analysis of change in the underdeveloped countries themselves, I hope this book will be of value to U.S. policy makers in India directly, and in other countries by analogy or use of similar methods. By relating changes in economic policy in India to political changes, this study will try to indicate various alternative development paths that are not only economically optimal or feasible, but that are politically possible as well—that is, where interests converge or diverge to make policies possible.

In order to understand the relationship between political and economic change in underdeveloped countries, the key fact to realize is that although these countries may be underdeveloped in an economic sense, they are frequently highly developed in the sense of having a well-established structure of political and social power—a structure that existed even under foreign rule. Economic development will almost inevitably upset this existing structure. This change may not occur initially or deliberately, but it will—as new groups [2] gain economically, or as there are shifts in relative economic position—in this sense: those groups that gain also seek to increase their political power, their status position, or both; and those groups that lose, either in an absolute or a relative sense, seek to protect their former position or minimize the shifts. This gives rise to political unrest and conflict, which may take a violent form; and although it will almost inevitably create serious problems for the foreign policy of the United States, it will also provide opportunities to influence those developments.

The process of development today is such as to heighten the political issues associated with the economic changes. First, the governments of many underdeveloped countries by their plans for development which determine the allocation of resources, by their own direct investment in particular sectors of the economy and particular areas, and by the controls—open or covert—they frequently place either upon private access to resources or private ability to engage directly in activities are deeply involved in economic life. Thus the development plans and policies become the very stuff of politics, with decisions being determined by political action, and the effects of those decisions in turn having profound political consequences for the party or administration responsible for the decisions.

[2] In Appendix A and Chapter 2 I shall attempt to define the groups I deal with precisely.

Second, today it is hoped that the process of economic development will be far more rapid than in the earlier centuries. Europe and Japan were already highly developed by the eighteenth century, in the sense that subsequent economic development could be built upon a set of existing social, political, and economic institutions that were conducive to further economic change and had been created over centuries. Many of these institutions—a unified country with a strong central government, a relatively high degree of literacy, a set of efficiently working economic institutions such as banks, roads, and so on—either do not exist in today's underdeveloped countries, or exist in a highly attenuated form. These countries are attempting to do in a few years for their entire societies what today's developed countries accomplished over centuries, and the wrench of the changes is therefore likely to be greater.

Finally, economic development today takes place under conditions of mass participation of the population in the process. In many countries there is a demand for economic improvement that extends throughout the entire population; furthermore the governments, even dictatorships, supposedly represent most of the population. Among the mass political movements the Communist movement is a force—legal or underground—one of whose main programs is economic change, which points to the examples of Soviet Russia and China as evidence that such change can be achieved most rapidly under communism, and which considers and advocates class struggle and overthrow of non-Communist governments as either a precondition of economic development or an inevitable result of it. In the eighteenth and nineteenth centuries the governments were limited in their popular support: they made little attempt to get mass support and had much less need for it than governments today; there was no general demand for economic improvement that had to be satisfied, and there was no influential Communist party.

Even though there were these differences between the conditions of growth that existed in the past and today's environment, the economic changes that occurred in the past were accompanied by profound political changes. It is unnecessary to try to determine which was the first cause since both were closely interwoven, but the economic changes undermined the remains of the European feudal system and were closely associated with the English revolutions of the seventeenth century, the French Revolution, the English unrest of the first half of the nineteenth century, and the European revolutions of 1848. Marx of course made this relationship the key to his theory of revolutionary change, but more conservative writers also recognized the importance of the interrelation. Lord Acton, writing about the French Revolution, states that "in attacking feudalism . . . the middle class designed to overthrow the condition of society which gave power as well as property to a favored minority. The assault on the restricted distribution of power involved an assault on the concentration or wealth. The connection of the two ideas is the secret motive of the Revolution." [3]

De Tocqueville described France on the eve of the revolution

> as a nation, within which the desire of making a fortune kept spreading day by day . . . with a government which continuously excited this passion, yet continuously harassed, inflamed, and drove the people to despair, thus thrusting on both sides towards its own ruin.

He contrasted France with England, where

> shutting your eyes to the old names and forms [which still are used], you will find from the seventeenth century the feudal system substantially abolished, classes which overlap, nobility of birth set on one side, aristocracy thrown open, wealth as the source of power, equality before the law, office open to all; . . . all new principles, of which the society of the Middle Ages knew nothing. Now these are just those novelties, which, introduced gradually and skillfully into the ancient framework, have reanimated it without risking its destruction, and have filled it with a fresh vigour, while retaining the ancient forms.[4]

Although De Tocqueville seems to have forgotten the seventeenth-century English revolutions, and did not consider the Chartist agitation of the 1840's, the contrast he cited is highly relevant. England was able to make the shift in political power associated with economic change in the nineteenth century relatively peacefully, in part because of deliberate policy that recognized the necessity, even if "undesirable," of changes and their proper timing. This policy reflected the past experience of the English landed aristocracy, which had retained political power for many centuries precisely because it was able to make the adjustments necessary to retain political power. This experience was lacking in France in 1789, and at a later date in Russia in 1917, and this lack contributed to the violent overthrow of the old regimes in both those countries.[5]

In Japan, somewhat as in England but for different reasons, the downfall of the Tokugawa Regime, which had ruled for about 250 years prior to 1865, reflected the burgeoning of new economic and political forces

---

[3] Lord Acton, *Lectures on the French Revolution* (New York: Noonday Press, 1959), p. 53.

[4] De Tocqueville, *L'Ancien Régime* (Oxford: Blackwell, 1947), pp. 189 and 21.

[5] On England, see J. H. Hexter, *Reappraisals in History* (London: Longmans, 1961), especially chapter II; on Russia, Trotzky, *History of the Russian Revolution*, Vol. I (London: Victor Gollancz, 1934), chaps. i–vii.

and the groups from which they arose. These combined with the demands of traditional social groups that had lost status under the Tokukawa Shoguns, and felt severely constrained by the old regime. But both the weakness of the old regime and the enlightened policy of the new regime of the Meiji resulted in a rather mild transition, which permitted the use of the long-developed institutions and traditions by the new ruling groups in favor of rapid growth policies.[6]

The experience of the United States has been sufficiently dissimilar, though related, to that of Europe to result in a different pattern of American history. However, even the difference confirms the general rule. In the United States there was no firmly established feudal order to be overthrown. Although the American Revolution had elements of group conflict, the structure of power and status that existed prior to the Revolution was not so established to call forth either the effort to overthrow it or the reaction to the overthrow that occurred in Europe. European immigrants to the United States in the seventeenth and eighteenth centuries were leaving the feudal system of Europe, and there were relatively free opportunities for jobs and ownership of land in the United States. These facts contributed to an agreement on the fundamentals of a social and political system based on equality of opportunity in the United States; consequently there has been a relative freedom from both extremes of social conflict that have characterized European history, and from the ideologies of such conflicts, especially Marxism. "It is not accidental that America which has uniquely lacked a feudal tradition has uniquely lacked also a socialist tradition. . . . America represents the liberal mechanism of Europe functioning without the European social antagonisms." [7] The peculiar American experience has colored the past American response to the process of change in the underdeveloped countries, and has affected American policies toward that process.

These few examples of the interaction of political and economic changes and the potential conflicts arising therefrom are mentioned to indicate, first, that economic development is an unsettling process even in countries functioning under relatively more favorable circumstances

---

[6] On Japan, see especially the two studies of Thomas C. Smith, *Political Change and Industrial Development in Japan: Government Enterprise, 1868–1880* (Stanford: Stanford University Press, 1955); and *The Agrarian Origins of Modern Japan,* (Stanford: Stanford University Press, 1959).

[7] Louis Hartz, *The Liberal Tradition in America* (New York: Harcourt, Brace, & Co., 1955), pp. 6, 16. This is not to say that the United States does not have its own types of conflict influenced by such characteristics of its history as slavery, which interestingly enough bears some resemblance to Indian caste conflicts, its tradition of the frontier, and the nineteenth- and twentieth-century economic problems that gave rise to Populism and the New Deal.

for peaceful change than today's underdeveloped countries. Second, deliberate policy applied at the appropriate points of tension and time can serve to dampen the potential violence associated with the rapid change.

India at present is going through an experience more like the European than the American. Even though India was under British rule for almost 200 years, which had profound effects upon the Indian social structure, the caste system was the existing social system in rural India at Independence. The caste system determined the ritual and social status of a group relative to other groups and of the individuals composing it, the economic functions of a group and the members composing it as well as its relations to other groups, and finally the political power of the group and its members.[8] In part as a result of the changes that occurred under British rule, the foundations of the caste system were gradually being eroded and being replaced by elements of a class system, especially in urban areas. Caste and class factors interacted to influence both the system of government under the British at local and national levels and the structure of power and ideas within the Congress movement, which led the revolt against the British. Independence, by its very character, profoundly changed the system of government and made it possible to introduce many "modern" ideas into legislation. Following Independence, the process of establishing a democratic government and political parties—as distinguished from a movement—and the reform legislation resulted in changes in the membership of the ruling Congress party, changes in the structure of intraparty power whether based on caste or class, and changes in ideas; and these changes in turn influenced the subsequent legislation adopted, the structure of the economic plans, the implementation of both the reform laws and plans, and thus the gains and losses to various groups from the economic plans and policies adopted. The interaction of these changes has contributed greatly to the fact that India has been successfully able to go through the past fifteen years peacefully, democratically, and fruitfully. In turn the interrelated series of past changes will influence future plans and actions, India's responses to the continuously arising problems of development, and thus the future economic and political development of India.

I have briefly summarized the Indian case, but this process is not peculiar to India among the newly independent countries. In many other countries there exist similar problems of the relationship and possible

[8] I will not get into the sterile argument of whether the Indian caste system and the related political systems prior to the British rule were forms of feudalism. The best analysis of this is D. Thorner's essay on India in *Feudalism in History*, ed. Rushton Coulborn (Princeton: Princeton University Press, 1956).

conflict between the position of those groups that wielded power in the predevelopment society, normally prior to Independence, and the groups whose power arises from the process of economic development, or increases because of it. These divergences have already contributed to the political conflicts and economic policy issues in the underdeveloped countries, and they may be expected to continue to do so in the future.

Political problems associated with economic change obviously raise challenges for the United States in the countries concerned. As a result of the U.S. political tradition of liberalism, with its relative absence of sharp group conflict based on economic changes, there has been some reluctance to recognize the existence of such basic conflicts arising from economic change in other countries;[9] this has been accompanied by an assumption that the conflict can be readily compromised. Both have contributed to a belief that economic development is normally associated with political stability or democracy and, as a policy implication, that economic aid itself is the major method to achieve political results beneficial to both the aided country and the United States. Fortunately this set of beliefs is now under examination, and there is already a wider range of policies and goals in Latin America as evidenced by the Alliance for Progress.[10]

The recognition of potential sources of group political conflict over economic development also opens opportunities for the United States that in the past have been ignored. United States policy tools, including aid programs and technical assistance programs,[11] often do benefit one or more groups in the aided countries and harm other groups; and the success of the programs or the acceptance of even ostensibly technical advice may depend upon a recognition of the groups that are aided or harmed, their place in the existing society, and the effect of U.S. policies upon them. Since American programs do influence the internal power structure, they may also be used consciously to affect the structure of power in the countries concerned and to influence groups within those countries in the direction of policies that contribute both to internally

---

[9] This may also be associated with some reluctance even to consider the parts of Marxist analysis as valuable for U.S. policy, simply because it is Marxist or Communist.

[10] For related treatments of these themes see Mancur Olson, "Rapid Growth as a Destabilizing Force," *Journal of Economic History,* XXIII, No. 4 (December, 1963), 529–552; and C. Wolf, Jr., "The Political Effects on Economic Programs," *Economic Development and Cultural Change,* XIV, No. 1 (October, 1965), pp. 1–20.

[11] There are of course different aid programs—economic and military—and a variety of policy tools other than aid—both diplomacy and trade policy—to achieve United States goals in dealing with other countries; and all, hopefully, have their effects.

desirable developments in the aided countries and to the associated interests of the United States. There has been some unwillingness to make conscious use of U.S. programs for such purposes. I think even if they are not so used they have political effects; and the undeliberate effects may be in directions that would be considered undesirable in terms of American goals. It is obvious that the Communist governments in Russia and China base their policies in the underdeveloped countries deliberately upon the recognition of group conflicts and attempt to appeal to groups and influence them; this has contributed to some of their past successes. Fortunately the rigidities of Marxist theory and practice have also contributed to serious weaknesses in Communist policy and subsequent setbacks. The United States does not have, or is often less willing to use, the instruments used by Communists to intervene overtly in the domestic policies of other countries. This is not necessarily a disadvantage to U.S. policy;[12] but the failure to identify contending groups within a country and to adjust policy means to influence results of the conflicts and in relation to the desired American goals has contributed to weakness and failure of American policy. At the same time it must be realized that U.S. influence and aid, however exercised and even at its greatest, may be only a minor factor in the political and economic developments within a country.[13]

In Appendix A, associated with this introductory section, I present the general framework of hypotheses within which the study proceeds and define the terms I shall use in applying this analysis to the interrelationship of political change and economic policy in India. Part II describes the well-established structure of Indian society at the time of Independence and prior to the economic development program. This section classifies the major economic and social groups within both rural and urban India at the time—emphasizing caste groups in rural India and class groups in urban India—and describes their economic and political positions, their goals, and their ideologies. It concludes with a chapter examining the influence of these groups upon the British government, and upon the Congress Independence Movement before 1947, in terms of the strength of their influence and as sources of leadership, ideas, and programs.

Following this rough picture of India in 1947, Part III describes the political changes arising from Independence and following from the

[12] This is a value judgment. I refer to the existence of Russian or Chinese financed Communist parties especially.

[13] For one of the best case studies of American policy in an underdeveloped country—one in which policy failed—I recommend Tang Tsou, *America's Failure in China, 1941–50* (Chicago: University of Chicago Press, 1963).

ideological and institutional changes associated with that upheaval. Appendix A to Part I presented an equilibrium theory of a social system; the political changes of Independence and thereafter upset the pre-1947 system, and by so doing led to the economic planning and policy changes of the 1950's. For this reason I discuss the changes in the structure of the Congress party in terms of the power of the groups identified in Part II, the ideas within the party, the role of political leadership, and the changing role of government—and the changing relations among the party, its leaders, and the government—during the 1950's, before I discuss economic policies and results.

The political section is followed by the economic section of the book, Part IV, which starts with an examination of the broad economic policies and results from 1947 to 1961, including the transition period and the First and Second Five Year Plan periods. The bulk of the economic section will be devoted to an effort to estimate, for this period, the economic gains and losses, both in quantitative and qualitative terms, to the various class and caste groups identified in Part II.

Part V discusses the future. I first explore the implications of the previous analysis in order to present a picture of India's current economic problems. Next I examine alternative Indian policies that might be adopted to deal with the economic problems in the light of the political changes that have occurred, and finally I look at alternative paths of India's future economic development in the light of the policy alternatives and I indicate the political implications of these alternatives.

Part VI briefly examines U.S. policies in relation to India in the light of the previous discussion of possible future Indian developments. After examining U.S. goals in India and the policy tools to achieve these goals, I present some suggestions for U.S. policies that might contribute to achieving American goals with the tools that appear usable in the light of the projected Indian developments.

The book concludes with a brief chapter in which I indicate certain generalizations on the relationship between political forms—especially the approach to democracy—and economic change, derived from the framework of analysis and from its application to India, that may be applicable to other underdeveloped countries now in the process of development.

PART II

India at Independence

[To] understand thoroughly the Revolution and its work, it was necessary to forget for one moment the France which we see, and to proceed to interrogate in its tomb the France which is no more. . . . But as to the manner of conducting affairs; as to the real working of institutions; as to the exact relation of class to class; as to the conditions and feelings of those classes which, buried beneath the dominant opinions and manners of the age, made themselves neither heard nor seen; we have only confused and often faulty ideas.

I have attempted to pierce to the heart of this 'old order,' so near to us in point of years but so completely hidden from us by the Revolution.

De Tocqueville, *L'Ancien Regime*

The soil grows castes; the machine makes classes.

M. Young, *The Rise of Meritocracy*

2

Rural Caste and Urban Class

THE RURAL AND URBAN SECTORS

The main apparent division in India, transcending the various other social, economic, and political groupings, is that between the rural and urban sectors. Although there are many interactions and influences from one sector to the other, the social system of the rural sector may be usefully considered to be the caste system. Even though it is in the process of change, caste underlies the economic, social, and political relationships among the peasants. In the urban sector many of the characteristics that identify the caste system in the village have broken down and have been replaced by a system based on factors much less related to family and inherited position. To identify and understand the position of the major groups in Indian society at Independence, major emphasis will be placed on caste groups modified by class elements in rural India, and on class groups modified by caste elements in urban India.

To distinguish between the rural and urban population, a crude line of demarcation, used in the 1951 and 1961 Indian Censuses, is that those inhabitants living in villages of approximately 5,000 population or less are considered rural population whereas those living in larger conglomerations are considered urban.[1] In 1951 there were approximately 559,000 villages in the rural category; most of these contained only about 300 inhabitants, and their total population was 295 million people. In contrast to the rural sector, there were 3,000 towns, each with 5,000 or more inhabitants, with a total population of 62 million. Thus about 80 per cent of India's population in 1951 lived in rural areas and 20 per cent in urban areas. Without entering into too much detail, the 80 per cent of the population in the rural sector produced approximately two-thirds of India's national domestic output in that year. The social system that governed the behavior of this rural population was the caste system.

RURAL CASTE

The caste system may be defined as the network of interrelationships among castes that governs the economic, political, and ritual behavior of

[1] The 1961 Census also introduced two other criteria: that the density be not less than 1000 per square mile, and that three-fourths of the working population be employed outside of agriculture. This led to the dropping of 812 places, and the inclusion of 483 new "urban" places from 1951 to 1961.

the castes within the system, and thus of the members of the castes.[2] Rather than attempting to define a caste in general terms, I will describe it by some of its broad characteristics.[3] The castes are ranked in a hierarchical order of ritual purity with the Brahmin castes universally ranked at the highest and the unclean castes at the lowest level. The ranking of the specific castes is traditionally associated with certain specific occupations, based in large part on the character of the materials handled and the functions performed in religious ceremonies; a caste seeks to protect its status and rank by defining behavior that is not permitted and punishing those caste members who perform acts that are not permitted, with the ultimate punishment that of outcasteing. The first major limitation of behavior is in connection with marriage: a caste member may marry only a member of his or her own caste excluding closely related family members as defined by the caste rules. A second major limitation is in terms of social contact, especially at meals or the occasion of physical contact with either members of the lower castes or the objects handled by lower castes; such contacts are carefully regulated and may occur only under prescribed conditions, and any violation results in ritual pollution and requires penance or punishment. Finally this system of hierarchy and controlled relationship is an integral part of the Hindu religion; it proceeds from its structure of reward and punishment, which directly relates the present caste position and future prospects of the caste member, in his reincarnation, and his descendants in the world to his ancestors' past obedience and his own present obedience to the prescribed rules of caste behavior.

It is important to realize that although the caste system functions throughout all rural India and national generalizations with respect to its functioning are possible, its detailed characteristics, especially in terms of ranking of castes between the Brahmins on top and unclean castes at the bottom, vary from area to area.[4] The basic caste unit within the village is not the broad caste division but the local subcaste or *jati*. A subcaste has relations—most probably for marriage purposes, but also for ritual and self-governing purposes—with subcaste members in other nearby villages, and ranking is similar in this area. The marriage radius for any one village subcaste is frequently ten to fifteen miles; but the network of marital and ritual relationships and accepted ranking order ex-

---

[2] I am excluding discussion of the family as part of the social system, although I recognize its major role within the caste structure. A caste may be considered as an extension of the family.

[3] The bulk of this description and the generalizations presented in this chapter are based on the very wide literature on caste, some of which I list in Appendix B.

[4] See the regional bibliography in Appendix B.

tends to some degree to all castes within the same language area. Language can be used to define community, so that within the same communal group there is more probably a similar caste structure and ranking than between different linguistic groups and areas.[5]

Although it has always been difficult to establish the ranking of a specific caste on a national basis because the local variations in ranking are so great, the fourfold national division of the clean castes—Brahmins, warriors, traders, peasants—and the unclean castes together provide a crude method of aggregating numbers on a national basis and also of making rough interregional comparisons. Moreover, the national agreement on the ritual importance of the Brahmins and on their role in the system contributes to making Hinduism a national religion, and also to an accepted body of theology and religious practices on a national basis. This has been an extremely important unifying factor in Indian history.

The traditional caste system not only provides a horizontal stratification of the elements in Indian rural society, for which it is best known, but also, and of equal importance, determines vertical relationships among the horizontal ranks. The caste system determines the economic functions of the various caste groups within the village, where the traditional relationship is that of patron to client (*jajman—kaman* in North India—so that in the economic sphere the functioning system is frequently called the *jajmani system*); this system also determines the distribution of the crops among the village members. In the ritual field the caste system prescribes relationships among all the castes; and each caste, even the lowliest, has definite functions that only its own members can perform in religious and social ceremonies. In both of these spheres every caste has some dependence on every other caste, and although it may be minimal, it is also essential. The Brahmin caste is the final decision-maker on ritual questions. In the purely political sphere—although there is generally intracaste government—the vertical relationship among the castes is far more complex and cannot simply be explained on ritual or economic grounds, although it is related to both. Prior to English rule, the ritual decisions made by the Brahmins were enforced by the political ruler and his legal system, regardless of the religion or the exact caste position of the ruler.

In Appendix A I consider a social system as an equilibrium system. The most useful theoretical concept of equilibrium in the caste system is that of dominant caste. Essentially, this states that when a particular

---

[5] I also define as communal, as distinguished from caste, those groups that are not of the Hindu religion—Parsees, Catholics, Moslems, British, Sikh, and various tribal groups—who are not united by language, but by their own religion or custom and by the fact that they are not Hindus.

caste enjoys a position of dominance in one sphere it tends to acquire, over time, a similar position of leadership in the other two spheres, so that it is dominant in all aspects of village life.

In the traditional economic sphere, in an agricultural and largely self-sufficient village, dominance is derived from the control of the use of village land. The members of the dominant caste in the village are the largest landholders, whether by conquest, original settlement, gift, or—since British rule—by purchase. Around these dominant caste members, who are often an extended family, there are other caste groups, in many cases with their occupations determined by tradition, performing the specific tasks necessary for village life. This includes the ritual functions that only Brahmins can perform; the actual work in the field, which the landholders are often not able to perform for caste reasons; the specific services necessary for production, consumption, or ritual ceremonies; and finally the occupations that revolve about the handling of refuse, dead matter, animal waste, and village sanitation. There may, of course, be some overlapping of functions, and the dominant caste may itself carry out certain of the other functions—as in a Brahmin village or, as in Mysore and Gujerat, where the cultivators are the dominant caste.

Under this system the members of the economically dominant caste are traditionally responsible for the output of the crop, and they thus control the income of the village. Each family of the landholding caste serves as a patron for a group of client families, which performs certain prescribed services for their patrons; that is, they perform ritual functions when necessary, work on the land as laborers, do carpentry, metalworking, and so on, and support the patron family in conflicts. To each of its clients the patron family assumes certain obligations with respect to employment, income, physical protection, and ritual services. The distribution of income among the members of the village is carried out at the sharing of the harvest, with the patron distributing the crop among the clients according to shares determined by custom and tradition and enforced by known mutual obligations. It is obvious that, were it in its pure form, there would be no distinction between caste status and class position in this economic system.

Ritual status centers about the caste group's role in interpreting the Hindu religion, and thus its purity and susceptibility to pollution. With respect to this scale, all other castes within the village are ranked in terms of their relation to the members of the Brahmin caste, although even here there are various grades of Brahminical purity. In parts of South India, one traditional measure of the ranking is the actual number of feet within which members of another caste can approach a Brahmin without polluting him. However, although the Brahmin caste everywhere is ritually dominant, it is much less often politically or economically dominant, since the Brahmin's traditional occupation is that of neither a warrior nor a cultivator of land. He is the custodian and interpreter of the Hindu religion and tradition and is necessary in the performance of certain sacred rituals. The Brahmin is required to know the Vedas, the sacred books, and probably is educated. Education has always been a source of wealth and power in India; before British rule a king or a local ruler would often select his advisor from the Sanskrit-knowing class, the Brahmins, and he would in return get religious support for his rule. In certain areas of India, especially in Madras and Kerala in the South, the ritual rank of the Brahmins in fact corresponds closely with their secular rank; they are dominant in all respects. In other areas of India this relationship is not precise, since the Brahmins are not the landholders. Nevertheless, since the presence and contribution of a Brahmin is necessary for almost all ritual services, the Brahmin caste has always had power in relation to other castes in the village, and this might be exerted as a last resort by a threat to withhold ritual services. But the weaker the Brahmins' economic position and the easier the opportunity to replace one Brahmin by another, the less the power is in fact. Normally a caste will try to solve intracaste difficulties by its own council, either on a village or area level. But if these are not so solvable, or if intercaste relations were involved, the disputants might appeal to a higher caste or to widely recognized spiritual and intellectual Brahmins, whose determinations help create the Hindu theology. Before British rule, both the Hindu and Moslem political leaders accepted the religious decisions of the caste leaders as law.

Political dominance is the power to make decisions within the village. The traditional institution of village government is the *panchayat*,[6] or council. With varying degrees of formality and overlapping of functions, such panchayats act to resolve disputes of various types that involve different groups of people—disputes within the caste; on an all-village level; on a level of patron-client economic relations; and, if need be, on a purely ad hoc basis for specific conflicts. For each of these there might be a different formal or informal panchayat. Normally, within a caste the older and more respected leaders are members of the panchayat; within the village, whether on a long term or ad hoc basis, the leaders of the caste panchayats—excluding unclean castes—will function as the village panchayat, but the wealthier members from the economically dominant caste exercise the greatest power. They are the ones who have more time to exercise leadership and the resources to perform the small favors

[6] I am treating here the traditional panchayat; not the statutory ones set up by both British and Indian law.

and cement the obligations necessary to achieve and maintain village support. At times these village leaders will even resolve intracaste disputes. The exercise of leadership in the traditional village panchayat context is somewhat different from that in the west; the panchayat leader aims at achieving a general consensus expressing the unanimous opinion of the village, rather than at being the voice of an individual representing himself or a majority. Where the dominant caste in the village is itself united, consensus can be readily reached; if it is split into factions, consensus becomes difficult and may be impossible. The decisions of the panchayat are enforced by religious, social, or economic sanctions, or a combination of these, depending upon the question and the need; but an element frequently present as the ultimate sanction—openly before the British and only less openly thereafter—has been physical force. As a result, caste or faction numbers, including retainers, were a significant element in political dominance before, although they have proved to be not as important since universal suffrage was introduced after Independence. Apart from the intravillage relationship, one of the major political functions of the village leadership is contact with the external world, especially the government which has the ability to give or take away so much and, apparently, so arbitrarily. Thus the ability to establish such contact, either by knowledge of a language or by literacy or education, or by other types of relationship, is a most important attribute of dominance. The intermediary role between the village and the outside government is a major source of political power within the village, and it is associated with the caste dominance.

There are a number of gains that the dominant caste achieves by virtue of its dominance, and that would be threatened by the loss of dominance.[7] To the extent that the dominant caste controls the village land, and that land is the source of all wealth and income in the village, the dominant caste can control not only the economic behavior but also the social behavior and attitudes of members of lower castes. In the economic sphere the dominant caste is assured a ready supply of cheap labor to perform all types of more or less menial and possibly polluting services—these included various forms of compulsory labor before 1947—ostensibly voluntarily, but with economic sanctions if the lower castes refused. In exchange the lower castes receive exclusive rights to certain polluting objects, and also a paternalistic care and protection, even in times of hardship for the village. But the dominant castes in many villages resent any effort of lower castes to get land or education, both of which would weaken the economic bases of dominance.

[7] Gerald D. Berreman, *Hindus of the Himalayas* (Berkeley and Los Angeles: University of California Press, 1963), pp. 242–254.

In the social sphere the members of the dominant caste have a gain in prestige and in feeling of superiority which is enforced by caste rules: the lower castes have to get out of the way of the upper castes; they are expected to be subservient to upper castes, and the upper castes are supported in their own superiority and judgment of their worth by this subservience. Another important social gain is sexual, derived from the access higher caste men have to lower caste women (but not the reverse, either in terms of sex or caste). Equally important in all areas, the dominant caste largely monopolizes access to the government, and thus controls the external sources of power and influence.

Finally, dominant caste members gain in opportunities for ultimate rewards in their afterlives—by their initially higher caste position, which encourages them to exercise their position in this life, and by their wealth, which permits them to give more to charity. Lower castes often do not have excess wealth to give to charity, and at the same time are enjoined by their caste rules not to strive for advancement on this earth, but to do their best to serve the higher castes. Thus there may be a conflict between the lower castes' hopes on this earth and their hopes in afterlife. (However, a member of any caste can escape the caste restrictions by becoming a *sanyasi*, a wandering holy man living on charity who has completely renounced the things of this world. This is the most looked-up-to role a Hindu can assume, and is open even to untouchables.)

Until now I have been describing the system as a static system. Obviously the caste system, as a very long-lived social system, must have provided opportunities for change and movement within it that did not require a break. The basic principle of change within the caste system was that the individual by himself is unable to improve his position relative to members of other castes within the system—although he may within a caste—but a caste may be able to improve its position, relative to other castes, and the individual benefits from the caste change. This principle is general for all castes except the Brahmin castes, which are already at the top of the ritual hierarchy and the unclean castes, which, as a result of their innate pollution, cannot raise themselves from unclean to clean within the system. This process of caste improvement is not a rapid one and may take a generation or more, but it is possible. The method of achieving this might be an initial improvement in one sphere: a relatively lowly caste migrates to a new area and becomes a landholding caste by settlement or conquest. With this improvement in one aspect of its position the caste *sanskritizes* [8] itself to improve its ritual status; that is, it adopts a new mythology appropriate to its hoped-for, higher status,

[8] This term to describe the process was first used by Professor M. N. Srinivas.

it finds a Brahmin willing to recognize its new status, it adopts ritual and dietary practices appropriate to its claimed higher status—it may abandon meat-eating since Brahmins frequently are vegetarian—and hopefully, depending upon its power or its geographic distance from its original location, it will eventually be recognized by other castes as having a higher status. The effect of this process is to reconcile shifts in political and economic power with shifts in ritual status, and this provides flexibility within the system. It is also an instance of the stable character of the system's equilibrium since the very process of sanskritization further strengthens the purity of the system's ritual forms and higher castes. However, sanskritization would not be possible for the unclean castes and would require acceptance by the Brahmins, who are already at the top.

Associated with the process of sanskritization as a dynamic element is the rise of factional groups within castes within a village. Even in the traditional caste system individual families were always changing position as a result of natural causes: disaster, the number or unbalanced sex ratio of their children, and the like.[9] As family positions improved or worsened there would be struggles within the caste; at the same time numerous opportunities would exist for conflicts within extended families over inheritance rights and other questions. The political form these conflicts frequently take is the creation of factions, with aggressive factions taking a more active role in sanskritization procedures.

In concluding this brief and very general description of a very complex phenomenon, I would like to emphasize again the fact that the caste system provides a remarkable example of a long-lived social organization. It performed many services. In association with Hinduism it provided a framework for stability of village life and custom that made it possible for the Indian village to exist as a unit in the periods of great political upheaval that have characterized so much of Indian history, and during periods of natural disaster that have been even more prevalent. At the same time it tied the village into a larger communal and national set of relationships, that provided a supravillage unity. The functioning of its institutions, such as the jajmani system and the panchayat, also contributed to patterns of behavior and a style that still strongly influence Indian economic and political practices and attitudes. However, although the system requires cooperation within prescribed rules, it has divisive as well as cohesive characteristics. First it is quite localized. A caste

[9] In the case of equal inheritance of property, obviously those families with more descendants to share might suffer; similarly, where dowries are given at marriage, families with a disproportionate female or male quota would lose or gain, depending on dowries paid or received.

member may know the caste of all the peasants in a fifteen-mile radius of his village: outside that radius, however, he may be confused and by making mistakes can harm his future prospects. The system thus encourages a perhaps natural tendency of suspicion among villagers. Within the village, both the caste system itself and the processes of sanskritization and faction-building, which so frequently accompany it, lead to a distrust of other villagers, friction with those outside the immediate family, and a constant jockeying for position.

> By and large the villagers rank their obligations to themselves and their families first, then to those fellow-caste members who live in the same village, and, at the end of the line, their fellow villagers and fellow castemen in other villages. This sense of obligation . . . does not extend to other villages as corporate bodies—much less to our vague notion of the "public interest" or "public weal" which embraces the whole population.[10]

And this leads to a profound distrust of outsiders or villagers with outside connections, or of the sincerity of those claiming to act for the general good or the public welfare, and not for themselves. But those few who, by their record and sacrifice, can convince the villager that they are acting selflessly have a great advantage in gaining and keeping the support of the peasant.

For the continued existence of the caste system in the form described above, a whole series of economic conditions were necessary. First, land had to be the main source of economic power, so that landholding insured economic dominance for the landholding castes. To the extent that relatively low-ranking client castes could acquire income from occupations not connected with land, they would thereby reduce their economic dependence upon their landholding patrons. Second, some equilibrium was needed between the numbers of patrons and clients within a single village and the presence of sufficient open land, such that an excess of either could migrate. For example, it is obvious that a small village could not support more than a few barbers; and if more barbers were born than could be supported, this would undermine the economic and social structure of the village. Similarly, if more new members of a landholder's family were born than the land could support in a customary style, this, too, would weaken the village social system. Both surpluses required some outlet, either in another village with a deficit of barbers or on a surplus of land held by the patron's caste members; more likely, the outlet was provided by open lands, which would permit the setting up of a new village that could support surplus cultivators and

---

[10] F. G. Bailey, *Politics and Social Change* (Berkeley and Los Angeles: University of California Press, 1963), p. 62.

barbers from the original one. A degree of flexibility was also provided by the fact that cultivation is not a closed occupation, so that surplus caste members of any clean caste could become cultivators on open lands. As population grew, these safety valves closed: the demand and supply relationship among patrons and clients in a single village became distorted, and the open lands were no longer available. Apart from the importance of the supply and demand relationship in determining the capacity of the village to support its caste system economically, the variations in numbers of patrons and clients also introduced the possibility of bargaining within the traditional limits. If there were only one barber in a village, for instance, he would be treated better by his patron families than if they had a choice of several barbers: in such a case he could gain more by hinting at departing. This reduced the severity of the system upon the lower castes. Finally, as a noneconomic requisite, the system's viability required that all the castes accept their positions, obligations, and duties within the system and the slow process of change it permitted.

All these conditions of the viability of the caste system were undermined by the fact of British rule from 1760 to 1947 and the developments during that period. First, the British, unlike previous rulers, did not accept caste decisions as a basis of law and in fact outlawed certain caste practices. Rather they set up their own administration and court systems, which were theoretically open to all Indians, regardless of caste, and to which anyone could appeal even caste decisions. Second, the British introduced a modified version of private property in land, which resulted in an extremely complex land system. It included, at one time, the right to individual purchase and sale of land, so that money income provided access to ownership.[11] Third, during this period the opportunities for earning money income from sources other than land increased greatly.

---

[11] The following quotation from D. Thorner, *The Agrarian Prospect in India,* (Delhi: University Press, 1956), chap. 1, provides a brief summary of the character of the British land reforms and the result:

> The key fact about all the British land settlements of the late eighteenth and nineteenth centuries . . . was that the new rights in the land were invariably subordinate to the rights of the State. To *no* holder was granted the exclusive right to occupy, enjoy, and dispose of land which, in *practice,* is the hallmark of Western private ownership. . . . What the British established in India might be described, in fact, as an imperfect or kaccha kind of private ownership of land. To this date there has not emerged in India a fully developed or pakka private property in agricultural land. Below the State's claim, elaborate sets of inferior claims were for the first time . . . put in writing . . . [and] we are all well acquainted with the layering of land which resulted . . . all over India . . . to a greater or lesser degree. Whatever the particular form of land system that was followed, it served to confirm the right of one group of holders to a share in the produce of the land. . . . [This right] tended to become increasingly separated from the actual tilling of the soil, and even from residence in the village.

As the evils of this system became apparent, legislation was introduced with

In part this arose from the peace provided by British rule; in part from the transport facilities that permitted a widening of the market; in part from the greater commercialization of agriculture and industry as the British and Indians entered trade for the purpose of profit from either the sale of goods to the cities or from exports to foreign countries; and in part from the new manufacturing industries set up by British and Indian businessmen, and from the new cities, which served as ports for trade and centers of industry. These new sources of income and wealth directly provided additional bases of economic power, and permitted lower castes that took advantage of them to buy land and become landholders. In fact, such land purchases were part of the sanskritization process for commercial castes, since land had so much more prestige than other types of wealth. At the same time, in the cities and new factories, many of the old caste restrictions were either impossible to observe or were not recognized. The British peace, combined with improved health and sanitation, contributed to a growing population; and although the annual rate of growth was small until about 1921, the more readily accessible open spaces became increasingly occupied, and the new opportunities existed not on the land but in factories and cities.

In the ideological field British education and British law regarded all men as equals; unclean castes were less willing to accept their low status once they realized it was changeable or they had a hope of improvement, and many broke away from caste rules. The Brahmins, especially, with their education, their experience in government, and their wealth, frequently sold land to become western educated, often adopted western ideas of equality, and then worked for the British or adopted western practices and behavior. In effect, for the Brahmin sanskritization meant westernization.

The British provided new chances for advancement in all castes to those members able to take advantage of them. For clean castes this could proceed through sanskritization; for unclean castes it would have to occur outside the system. But there were also more chances for decline: the breakdown of communal responsibilities for village members, which accompanied the introduction of private property in land; the breakdown of customary village occupations and services, which accompanied the

---

the aim of restoring the traditional rights of the actual cultivators. Largely this legislation took the form of tenant protection [confirming the rights of tenants, or forbidding further sale of land and dispossession]. . . . The result has been a layering of rights from those of the State . . . down through those of the sub-landlords to those of [the] several tiers of tenants . . . [each] collecting rents from the [group] with rights inferior to [its] own [down to the working cultivators]. . . .

Hereafter when I use the term landowner, I use it with the qualifications in this footnote, since it is a very complex concept in India.

spread of the market place and urbanization; the abolition of slavery for the lower castes, especially in the south—all contributed to the rise of a large group of landless laborers. These were usually of the lower castes, who now received not an annual share of the village's output as a matter of custom but only a daily wage when their services were required on the land, and they frequently sought additional earnings in the new cities. In almost all cases, both for high and low castes, the relation between a traditional occupation and caste either greatly widened to include not just a single occupation but a whole range of work or broke down entirely. Thus at both ends of the caste hierarchy, the ideology and practice of caste were under question; unclean castes were less willing to accept the sanctions of the system, since the system's support for them weakened, and those at the top were less willing to provide or use these sanctions.

But while these developments occurred, the British themselves did not welcome caste or communal social and political change at a speed that would cause political problems for them, and they were in fact willing to use some of the group differences to strengthen their position; at the same time the rate of economic and political change was not so rapid but that the caste system could more or less adjust to the changes in all spheres. Although the caste system was challenged and its underpinnings at all levels were being eroded, in fact the system functioned in rural India with now a rough and blurred correspondence of power and status in the spheres of ritual status, economic power, and political power.[12]

Table 1 summarizes the numbers and positions of four major caste groupings in rural India, and Table 2 relates these caste groupings to rural occupations for the same period, 1952–1954. The National Sample Survey of 1952–1954 presented total caste data for rural and urban households, based on samples, and divided the household population into four caste-rank groups—upper, middle, lower, and the scheduled castes.[13] The upper castes were defined as those who, according to custom, used the sacred thread, the middle as those from whom the Brahmins take water by tradition, and the lower as the other castes who were not scheduled.[14] *Scheduled castes* are the "unclean" castes listed in the Constitution, and receive certain constitutional privileges. Based upon this ritual ranking

[12] Almost all of the anthropological studies of the village caste system, on which the theoretical concepts of dominance and sanskritization are based and from which my generalizations are formed, were conducted in the 1950's.

[13] This data, from the National Sample Survey of 1952–1954, is the only fairly recent national data on caste. The Censuses do not ask this, nor have later National Sample Surveys.

[14] The sacred thread is worn by the three leading caste groups, and is assumed at a very important ceremony at a certain age.

—determined by relation to Brahmins, which is not as precise as it seems since many castes have been assuming the sacred thread—the survey divided the total household population of rural India into one of the four above caste groups as of 1952–1954 and related this ranking to the group's occupational function.[15] K. N. Raj, in a 1961 article, conveniently summarized these data, and Tables 1 and 2 are derived from his summary.[16]

Although there is a good deal of potential overlapping both in caste ranking and in occupational definition[17] these two tables provide useful information: they summarize the only recent and available data on caste numbers, both nationally and by regions. Thus, although the upper caste groups consisted of less than 10 per cent of the total rural household population, they were found to be of relatively greater importance in the north, east, and northwest regions, and to be of very small numerical significance in the south, west, and central regions. This may also imply that caste lines are stricter and more clear cut in the latter regions; whereas in the former there may be more crossing over the lines and blurring them as lower castes assumed characteristics, such as the sacred thread, of higher castes and reply accordingly.[18] At the other end of the scale, although the scheduled caste households were more or less evenly distributed throughout rural India, the lower nonscheduled caste households were disproportionately concentrated in the south, west, and central portions of India. These absolute numbers have proved to be of greater political consequence since Independence and the introduction of democracy; they were much less important in determining group power under British rule.

There was also, on an overall basis, a positive relationship between caste rank and occupational status, especially at the extremes. Over 90 per cent of the upper caste households employed in agriculture were farmers and cultivators, whereas almost 50 per cent of the scheduled caste households in agriculture were agricultural laborers, and another

---

[15] GOI Cabinet Secretariat, "National Sample Survey," No. 14 (Calcutta: 1958), p. 17. See K. N. Raj, "Regional and Caste Factors in India's Development" in *Tensions of Economic Development in Southeast Asia,* ed. J. C. Daruvala (Bombay: Allied Publishers, 1961), pp. 108–113.

[16] *Ibid.,* p. 110 for Table 1 and p. 112 for Table 2 below.

[17] Raj rightly makes the point that there is a substantial overlap between the classes of cultivators and sharecroppers.

[18] The introduction of lists of backward classes and reservations for those classes in education, government appointments, etc., may encourage households to reply as lower castes, especially in the south where there is also a good deal of anti-Brahmin feeling. Also certain castes may simply not be present, or only in very small numbers, in some areas—for example, upper class Vaishyas in West Bengal in eastern India.

TABLE 1
DISTRIBUTION OF RURAL HINDU HOUSEHOLDS BY MAJOR CASTE RANKS, 1952–1954
(millions)

Rural Areas of India	Total Number of Households	Total Number of Hindu Households	Hindu Households by Caste Ranking[a]			
			Upper	Middle	Lower	Scheduled
North	10.9	9.7	1.8	3.7	1.8	2.4
	(100.0)[b]	(100.0)	(18.3)	(39.1)	(18.5)	(25.1)
East	16.6	14.6	1.6	5.0	4.4	3.7
		(100.0)	(10.8)	(33.9)	(29.8)	(25.5)
South	12.6	10.6	0.3	0.5	7.5	2.4
		(100.0)	(2.6)	(4.5)	(70.9)	(22.0)
West	5.6	5.2	0.1	0.5	4.0	0.6
		(100.0)	(2.2)	(10.1)	(76.2)	(11.6)
Central	9.3	8.8	0.3	1.1	5.9	1.6
		(100.0)	(3.2)	(11.9)	(66.3)	(18.6)
Northwest	5.2	4.2	0.5	1.6	1.1	1.1
		(100.0)	(11.3)	(38.7)	(24.6)	(25.4)
Total	60.1	52.9	4.5	12.2	24.4	11.8
		(100.0)	(8.4)	(23.1)	(46.2)	(22.3)

Notes:
[a] The figures in millions have been rounded to one decimal after computation.
[b] Figures in parentheses are percentages.
SOURCES:
K. N. Raj, "Regional and Caste Factors in India's Development," in *Tensions of Economic Development in South-East Asia*, ed. J. C. Daruvala (Bombay: Allied Publishers Private Limited, 1961), p. 110; National Sample Survey No. 14, issued by the Cabinet Secretariat: Government of India (Calcutta: Eka Press, 1958–1959), p. 28.

TABLE 2
OCCUPATIONAL DISTRIBUTION OF DIFFERENT CASTE RANKS AMONG HINDU RURAL HOUSEHOLDS, 1952–1954
(millions)

Occupation [a]	Hindu Households by Caste Groups [b]				
	Upper	Middle	Lower	Scheduled	Total
Agriculture					
Farmer	1.1	0.9	1.7	0.2	3.8
	(24.4)[b]	(7.6)	(7.0)	(1.5)	(7.5)
Cultivator	2.0	6.5	10.2	3.2	21.9
	(43.9)	(53.3)	(41.8)	(27.1)	(41.4)
Sharecropper	0.2	0.8	1.5	1.0	3.4
	(3.9)	(6.2)	(6.2)	(8.5)	(6.5)
Agricultural laborer	0.1	1.5	4.1	4.3	9.9
	(1.1)	(11.9)	(16.8)	(36.2)	(18.4)
Forestry, fishery,	0.02	0.2	0.8	0.3	1.3
& livestock [c]	(0.6)	(1.4)	(3.3)	(2.3)	(2.4)
Total, agriculture	3.3	9.8	18.3	8.9	40.4
	(73.9)	(80.4)	(75.1)	(75.7)	(76.3)
Others [d]	1.2	2.4	6.1	2.9	12.5
	(26.1)	(19.6)	(24.9)	(24.3)	(23.7)
Total	4.5	12.2	24.4	11.8	52.9
	(100.0)	(100.0)	(100.0)	(100.0)	(100.0)

Notes:

[a] The main farming occupations are classified into four groups: (a) *Farmer*—a tiller who cultivates his own land, mainly with hired labor; (b) *Cultivator*—one who cultivates land mainly owned by him and sometimes land taken on lease or sharecropping system, with the help of other household members and partly with hired labor; (c) *Sharecropper*—one who mainly takes up cultivation of others' land on a sharecropping basis and cultivates without hired labor; and (d) *Agricultural Laborer*—one who cultivates others' land either for wages or for a customary payment.

[b] Both the figures in millions and the percentages, in parentheses, have been rounded to one decimal.

[c] Includes wood cutters, plantation labor, gardeners, fishermen, animal breeders, cattle grazers, and herdsmen.

[d] Includes households in the rural sector engaged in administrative and professional services, teaching and medicine, manufacturing—especially of food products and textiles —trade and commerce, transport and communication, construction and sanitation, and mining.

SOURCE:
N.S.S. No. 14, pp. 28–29, 223–224, and Table 1.

10–15 per cent were sharecroppers or engaged in forestry and animal occupations. With respect to the lower nonscheduled castes, approximately one-third of the agricultural households were employed as sharecroppers and laborers.

It is clear that, at least on a nationwide basis, by 1952 there was a good deal of overlapping between apparent caste and occupational rankings. Although only a relatively small proportion of the lower clean caste households were farmers, in absolute terms the number of such low-caste farmer households exceeded that of any other caste group; and the same caste group had the largest numbers of households in the second-rank cultivator occupation. Unfortunately, the relationship between caste-group and occupation was not broken down by region—it is known, for example, that in Mysore and Maharashtra many of the landowners are not upper caste—nor was any attempt made to relate either size of landholding or amount of land cultivated by the farmer, cultivator, and sharecropper groups to caste ranking. The latter relationship may, of course, be far more important than just the physical fact of owning some land.[19]

The Report of the All-India Rural Credit Survey summarized what had happened to caste power in land under the new British property laws and with the new opportunities for income:

> The bigger landlord has ways which conform with those of the moneylender, and indeed, as we have said, he is often the moneylender or trader himself. The village headman is also drawn from the same class, and it is usual for these to have connexions which link them . . . to the seats of administrative power. . . . The rigidity of caste loyalty remains, while the original division of caste functions no longer does. The result is that the landlord who may also be moneylender, the moneylender who may also be trader and the educated person who may also be subordinate official, all these through their association with the outside urban world of finance and power wield an influence in the village which at many points is diverted . . . to the benefit of the caste or even a close circle of relatives.[20]

The new claims of the clean intermediate and upper castes might have been fitted into a new equilibrium of class and caste position in the system; but the claims of the untouchables create the indigestible social lump within the caste system. It is their still-limited claims for economic development, for political power to go with their numbers, and for social power to go with their expectations—all of which are supported by urban and intellectual pressures toward equality—which are opposed most strongly by the dominant castes within the village who see themselves threatened. For this reason some untouchables under the British began to detach themselves from the limits of village society and appealed to British law; and since Independence they have begun to function on a political level in the wider regional or national arenas.

RURAL CLASS

A measure of class position in the land at the end of British rule is supplied by some of the scanty data on land holding summarized in Table

---

[19] We leave any discussion of the relationship of caste in the "other" classification of occupations to later, although it is of interest.

[20] The Committee of Direction, All-India Rural Credit Survey, *The General Report,* abr. (Bombay: Reserve Bank of India, 1955).

3. This table, derived from the pioneer work of D. Narain, presents data on the distribution of landholdings, in terms of area operated, by agricultural population and gross value of agricultural output and marketed surplus in 1950/1951; that is, before the major land reform laws were passed or became effective.

Table 3 shows rather clearly that by 1950/1951 approximately 30 per cent of the total Indian agricultural output was being marketed. Thus to the extent that a very high degree of village self-sufficiency was a condition for the viability of the caste system, that condition had been somewhat undermined. It also makes two very significant points concerning both the structure of landholding and the structure of market supply.

The 75 per cent of the agricultural population that held less than 10 acres of land worked about 40 per cent of the total acreage, produced approximately 50 per cent of the gross farm output, and supplied about 45 per cent of the marketed surplus. Table 3 does not consider the further fragmentation within the holdings, which is so characteristic of Indian villages. If measured this would still further reduce the average size of the plot worked to well below that of the holding. At the other end of the scale, less than 10 per cent of the total agricultural population owned holdings of 25 acres or more, but the operators of these holdings farmed another 40 per cent of the acreage, produced over 20 per cent of the total output, and supplied almost one-third of the total marketed surplus.

Second, the ratio of the part marketed to the total output was approximately the average for the smallest size holding, below five acres; it then declined sharply for holdings of 5–15 acres; this ratio then rose to about the average again for holdings of 15–25 acres; and it gradually rose to significantly above the average figure for holdings of 25 acres and more. Narain explains this pattern by the hypothesis that the cultivator of the smallest holding is a subsistence cultivator who sells under pressure to make certain required cash payments—for such fixed charges as taxes, rent, and debt; for necessities, such as textiles, salt, kerosene; and for various social expenditures. As the size of holdings increases up to 15 acres, these forced-sale pressures are reduced so that the proportion marketed is also reduced; beyond 15 acres there is a surplus above consumption which comes to the market or is held in stock for purely profit-making purposes.[21]

[21] It may be highly significant for agricultural policy that almost half of the marketed product comes from small holdings below ten acres, probably from forced sales. Thus, a small increase in output from these holdings if prices do not fall, or a relative rise in price with constant output, may lead to a reduction rather than an increase in marketed output.

TABLE 3

DISTRIBUTION OF OPERATIONAL HOLDINGS IN INDIA BY SIZE GROUPS, 1950–1951

Size of Holding [a] (acres)	All India Area Operated (thousand acres)	Agricultural Population, 1950/51 (millions)	Value of Output, 1950/51 [b] Total Gross (million rupees)	Per Cent of Total	Marketed Surplus (million rupees)	Per Cent of Total	Marketed Surplus as Per Cent of Gross Output
0–5	58,497	129.0 [c]	16,786	25.8	5,640	26.0	33.6
5–10	68,813	52.6	16,260	25.0	4,448	20.5	27.4
10–15	44,120	20.9	7,355	11.3	1,701	7.9	23.1
15–20	37,568	13.4	5,737	8.8	1,728	8.0	30.1
20–25	24,594	7.2	3,449	5.3	1,110	5.1	32.2
25–35	21,369	5.1	2,939	4.5	1,168	5.4	39.7
30–40	27,566	5.0	3,504	5.4	1,396	6.4	39.8
40–50	17,198	2.5	2,321	3.6	1,078	5.0	46.4
50+	50,600	4.9	6,614	10.2	3,399	15.7	51.4
Total	350,325	240.6	64,965	100	21,668	100	33.4

Notes:

[a] This is an operational holding, including nonagricultural lands, but this latter inclusion is felt to be factually unimportant.

[b] From estimate by D. Narain, which he considers superior.

[c] Includes agricultural laborers without land. In 1950/1951 the Second Agricultural Labor Inquiry estimated there were 9.0 million such households with a population of about 45 million.

SOURCES:

D. Narain, *Distribution of the Marketed Surplus of Agricultural Produce by Size of Holding in India, 1950–51* (New Delhi: Asia Publishing House, 1961), Statistical Appendix, Tables 1 and 7 (II) on pp. 39 and 44.

Thus with respect to the class structure of agriculture, it is possible to distinguish three groups in 1950/1951, before land reform: (1) The larger landholders, owning more than 15 acres and including a relatively few with more than 50 acres. This class, numerically somewhat small, was engaged to some extent in commercial farming and was interested in profitable operations. In 1950 it would have included princely rulers and many absentee landowners who did not operate their land for either caste or social reasons. It is this group that probably contained members of the then dominant castes, including the trader and landowner castes. (2) The second group consisted of the small landowners owning from five to fifteen acres. Their production was largely for subsistence; they were frequently hampered by lack of capital and other resources for investment or expansion. They were productive, with their lands frequently yielding a greater gross output per acre than from the larger holdings. Also, they were numerically much more important in rural India than the previous group. In many cases they were members of the higher castes; also, they were very frequently members of lower castes—clean castes, as well as unclean castes—who had been able to raise themselves into a better economic position and had purchased land as a result. They probably were aware, or might become aware, of new economic, social, and political opportunities. (3) The third group included groups with very small landholdings and various landless groups besides low caste landless laborers, village artisans and petty village traders, who were in many cases the clients of the dominant caste. Some had achieved petty landholdings or had moved down into this class as a result of continued subdivision of larger plots. Members of this class might have resources to seek other rural or urban employment, but they might be so depressed in all respects that they had no incentives for economic improvement or for attempting to raise their political and social positions, whether within or outside the caste system.

URBAN CASTE[22]

In urban India caste is increasingly being replaced by individual and class relationships. There is no clean break between the caste system in rural India and class in urban India, but rather there is a rural-urban continuum, in which the caste system still influences urban life. However, the manifestations of caste in urban areas take different forms from those in the rural setting. Class factors based on the role of an individual in the economy and society interact with caste elements. The caste and

[22] For the published literature on which this is based, see Section III of Appendix B.

class factors both support and oppose each other, but class elements are more important than caste elements in understanding urban society and pressures.

The caste elements enter from a variety of factors. First, the Indians living in the cities, in almost all cases, came from a rural background of caste. The urban caste system now operates in a new context; hiring by the government and by British commercial firms is not determined by caste but by the competence of the individual. However, some castes are in a better position than others to perform the functions or to learn the skills required by the government or private commercial and industrial firms; and these caste groups tended to concentrate in certain types of occupations. Also, various cities have attracted different caste groups, and it is difficult to determine a clear national pattern between caste position and migration. Brahmins generally have been attracted to urban white-collar work. In Kanpur, where the leather industry is a major employer, many unclean caste members, who are ritually allowed to handle hides, have moved into the city, and some have done well; in more recent years higher caste Hindus have come to Kanpur in increasing numbers, and some have even worked in the leather industry, although few handle hides. In Mysore state, some members of the dominant caste, who are on the margin in terms of landholdings, have taken urban jobs, but consider such a move temporary to build up their resources for purchases of more land. In Calcutta many male members of small landowning families of cultivating castes and members of artisan castes have moved to the city, and frequently they send remittances back to their families in the villages. In fact, movement to the city implies a minimal level of finance and a minimal number of contacts by the movers, which the lowest castes frequently do not have unless their skills are especially desired.

The urban areas are varied in their exposure to western influence. Cities have always existed in India as religious and political centers, and these older cities reflect the caste structure of their area. Other cities—especially the four great contemporary centers of Calcutta, Bombay, New Delhi, and Madras—were built by the British as trading and ruling centers, and these have never been based on caste, but rather have attracted people from all India and of all castes and communities. In some of the older cities and in cities that have grown from traditional roots, such as Madurai in the south, much of the city's land is owned by members of the regional dominant caste, and the members play a major role in the city's economic, political, and social life. But in such cities, too, the dominant castes have either changed over time or have assumed trading, landowning, and moneylending functions, and thus have become classes rather than castes.

In the large cities especially, many of the ritual elements of rural caste could no longer be maintained. In factories and buses it was impossible to preserve strict rules against physical contact; with the shortage of housing some caste intermixture in housing colonies became necessary. The one element of village caste that was more or less maintained was the rule against caste intermarriage, and many urban marriages are still made only within families and castes from the same village. In the economic sphere the breakdown of traditional caste occupations has gone far, but castes have sought to preserve their general type of work. With their financial resources, background of education, and tradition of advising rulers, Brahmins, under the British, were among the first caste groups to take advantage of urban opportunities, especially in the areas of government, the professions, and clerical work—and they reaped a harvest of dislike. Brahmins rarely entered business; the merchant castes, with experience in money-making, played a prominent role in trade and commerce, and have taken the lead in entering industry since the First World War. However, very little detailed research is available on the part that caste factors play in urban life: although impressionistic information is available, few studies comparable to the detailed village studies of caste functioning have been carried out; and the impressions are frequently confusing, if not directly contradictory. Tables 4 and 5 summarize the caste position of India's economically active Hindu urban population in 1952–1954 on a national basis.

It is possible from Table 4 and the total figures of urban households given in the final row of Table 5 to compute the absolute numbers of the various caste groups in various types of urban occupations.

Tables 4 and 5 show a tendency, as of 1952–1954, for a relatively greater proportion of the upper and middle caste households to reside in urban areas, whereas the lower castes, especially scheduled castes, tend to remain in the villages. They also show a very striking related concentration of the upper castes in administrative and professional service occupations, which are normally upper class white-collar occupations; the middle castes were also relatively concentrated in similar white-collar occupations in administrative and professional fields and in trade and commerce; whereas the lower and scheduled castes tend to be employed in greater proportion in the relatively lower class of product-handling occupations—agriculture, manufacturing, transport and communications, and construction and sanitation. This distribution reflects the greater access that the upper castes had to the western education necessary for white-collar jobs, because of their wider social contacts and economic advantages; and it demonstrates the prior entry of these castes into such jobs as part of the process of westernization. On the other hand, it reflects the traditionally higher status of white-collar jobs among the upper castes,

TABLE 4
PERCENTAGE DISTRIBUTION OF URBAN HOUSEHOLDS IN MAJOR HOUSEHOLD OCCUPATION GROUPS BY MAJOR HINDU CASTE GROUPS, 1952–1954

	Hindu Caste Group				
Household Occupation	Upper	Middle	Lower	Scheduled	All
Agricultural occupations	7.9	12.7	19.4	10.8	14.0
Nonagricultural					
Mining	—	0.2	—	0.2	0.1
Manufacturing (all)	5.2	11.8	14.1	16.4	12.1
Construction and sanitation	1.7	4.0	5.0	15.9	5.6
Trade and commerce	10.3	18.4	12.6	6.3	13.2
Transport and communication	7.7	8.2	10.8	14.2	9.9
Administrative and professional services	58.0	37.5	28.8	27.0	36.5
Others	9.2	7.1	9.3	9.2	8.6
All nonagricultural	92.1	87.3	80.6	89.2	86.0
Total of agricultural and nonagricultural occupations	100.0	100.0	100.0	100.0	100.0
Distribution of urban Hindu household population by caste:	17.4	29.5	39.4	13.6	100.0
Distribution of total Hindu population by caste	9.9	24.2	45.1	20.9	100.0

Note:
— indicates negligible.
SOURCE:
N.S.S. No. 14, p. 30.

and their unwillingness to take manual jobs; so that caste tends to support class, and vice versa.

These same tables show a substantial overlapping between occupational grouping and caste groupings for urban India. In fact the largest absolute number of people in the white-collar administrative jobs were from lower castes. Thus, although members of the lower castes might have been leaving their traditional occupations and going to the cities less readily than the upper castes, they were in fact doing so. In the process some of the members of the lower caste groups were entering middle-class and white-collar occupations; for in the urban areas, especially the larger cities, greater individual mobility was possible than in the village. Similarly, in the cities themselves lower caste organizations were embarking on political and educational programs which would permit their members to achieve a higher class status; and in all urban areas, but in the largest especially, class had probably become more important than caste in determining an individual's social position and political influence.

TABLE 5
ABSOLUTE NUMBER OF URBAN HOUSEHOLDS IN MAJOR HOUSEHOLD OCCUPATION GROUPS BY MAJOR HINDU CASTE GROUPS, 1952–1954
(millions)

Household Occupation	Hindu Caste Group				
	Upper	Middle	Lower	Scheduled	All
Agricultural	0.1	0.4	0.8	0.2	1.4
Nonagricultural					
Mining	—	—	—	—	—
Manufacturing	0.1	0.4	0.6	0.2	1.2
Construction, etc.	—	0.1	0.2	0.2	0.6
Trade and commerce	0.2	0.6	0.5	0.1	1.3
Transportation, etc.	0.1	0.2	0.4	0.2	1.0
Administration, etc.	1.0	1.1	1.2	0.4	3.7
Others	0.2	0.2	0.4	0.1	0.9
All nonagricultural	1.7	2.6	3.2	1.2	8.8
Total all occupations [a]	1.8	3.0	4.0	1.4	10.2

Notes:
— indicates negligible.
[a] Differences between sums and total of individual figures due to rounding.
SOURCES:
N.S.S. No. 14, p. 30; K. N. Raj, "Regional and Caste Factors in India's Development," Table IV, p. 110.

Although caste of the village type apparently declined in importance, communal and broader caste differences became more important. The leading business groups in the cities were communal groups: the British in Calcutta; the Parsees and Gujeratis in Bombay and Ahmedabad especially; the Marwaris throughout India, but especially in Calcutta; and the Sikhs in many areas of northwest India. Frequently these groups set up their own chambers of commerce, which functioned separately from other groups. The workers in the factories tended to come from certain rural districts, in part because they were recruited by jobbers coming from these rural areas rather than directly by the factories. In Bombay the factory workers often came from Maharashtra; in Calcutta from Bihar, Orissa, and Uttar Pradesh. The clerks were often high-caste Bengalis, South Indians, or Christians. The members of the same community lived in the same districts in the city; they celebrated festivals together, supported each other politically, and frequently set up educational, financial, and charitable institutions for their members in the neighborhood or city.

Under the British, caste was made a criterion for certain privileges; it became important for castes to protect or advance their status and to deal with the British government. To carry out this task, caste associations were set up that combined features of western voluntary organizations with their own literate bureaucracies and Indian caste organizations, which were restricted to members who claimed to be of the same broad caste group. These associations were usually in the same linguistic region, transcending a single village or group of villages. They were in part lobbying organizations, in part protective organizations, and in part educational and behavior-setting organizations. They functioned primarily in an urban social setting, dealing with the British government and the institutions set up by that government.

These broader caste and communal groups at times reflected class differences. For example: in Bombay, where the Maharashtrans tended to be factory workers, the Gujeratis and Parsees were the businessmen. In such a case, differences among caste or communal groups were reinforced by class factors, and vice versa, and the total effects of differences were multiplied. But there are many cases where the caste and communal factors, reflecting the rural background of many of the urban inhabitants, cut across class rivalries. For example: Hindu speaking factory workers from rural areas are a main support of the Congress party in Calcutta. Thus each has mitigated the full force of the other as divisive factors, and thereby contributed to a unity overriding either narrow communal or class lines. Ethnic factors are important in urban life: but the effect of a widening economic, intellectual, and political life which occurs in an urban environment; the range of interests, much wider than caste or community on the part of urban businessmen, intellectuals, and workers; the whole set of new functions and roles that are required in an urban environment; the western education found in urban areas—all have weakened the ethnic elements.[23] And in urban India the most important criterion of an individual's position has shifted from his membership in a caste to his own achievement and function in some area of urban life—to him as a member of a class rather than a caste or communal group. As members of a class, their economic and political interests frequently override the influence of narrower caste and communal attitudes. Thus, in urban India under the British, there was a rise of both middle and working classes. By the time of Independence these classes contained relatively large numbers and exercised both economic and political influence.

URBAN CLASS

THE URBAN MIDDLE CLASS

Any attempt to define the middle class can easily become a morass.[24] Instead it is far more useful for the purposes of this study to indicate the

---

[23] Following Weiner, I use "ethnic" as a broader term including both communal and caste factors.

[24] See on this subject J. H. Hexter, *Reappraisals in History*, chap. 5.

groups included within this class. On a top level it is clear that the leading businessmen in India, the leading members of the government bureaucracy, and the leading intellectuals should be included. The higher income professionals, scientists, and technicians, the professional managers in industry, and the large merchants should also be associated with the upper group within the middle class. Below these are the mass of clerks and relatively minor officials in government offices and private commercial and industrial firms, the school teachers, working journalists, struggling professionals, and the petty shopkeepers and possibly small-scale industrialists. Although this lower middle-class group generally has a high rate of literacy and education and an inclination toward white-collar jobs, individual incomes are low.

What is the total size of the urban middle class? As a rough total figure, it was estimated (see above, Table 4) that in 1952–1954 approximately 50 per cent of the 13 million urban households in India derived their main income from jobs in the categories of "trade and commerce" and "administrative and professional services." If it is assumed that these activities include all middle-class occupations, there would have been about 6.5 million urban middle class households in India in 1952–1954 (or less than 10 per cent of the 73 million households in all India).[25]

A finer breakdown by occupations is possible, and it will be useful even though it is crude. The National Sample Survey estimated that 36 per cent of India's urban population of approximately 66 million people in 1953–1954 [26] were in the urban labor force, or a total of about 24 million people; but about 950,000 were unemployed, and a substantial number were in agricultural occupations in urban areas. The last have been excluded from the urban class figures in Tables 6 and 7. Table 6 presents a partial occupational breakdown of the middle class, based on National Sample Survey No. 14; also it includes a somewhat more detailed, though still partial, breakdown of the administrative and technical category.

Within this total labor force figure of about 6.5 million members of the middle class, the upper income group in the early 1950's may be assumed to have been roughly equal to the number of individuals assessed for income tax on annual earned income above Rs 12,500 (or $2,500). In 1952/1953 this group numbered 65,000–70,000 assessees paying income

[25] "National Sample Survey," No. 14, pp. 29-30; K. N. Raj, "Regional and Caste Factors in India's Development," p. 110. Approximately 12 per cent of the 60 million rural households were primarily engaged in those classes of activities also.

[26] The 1951 Census figure was 62 million and the 1961 figure 79 million. This was a rise of 28 per cent; the 1953/1954 text figure of 66 million is an interpolated estimate, on the assumption that about one-fourth of the increase occurred between 1951 and 1953/1954.

TABLE 6

BREAKDOWN OF URBAN MIDDLE CLASS OCCUPATIONS IN INDIA, EARLY 1950'S

Occupational Group	Per cent of Labor Force		Number of People in Labor Force (thousands)	
Employers [a]	0.28 (of all urban population)		200	
Merchants [a]	10.0		2,300	
Retailers		7.0		1,600
Wholesalers, etc.		3.0		700
Professional employees			770	
Medical and health services [a]	1.0			230
Teachers [a]	1.7			390
Journalists [b]				100
Lawyers, n.e.c.[c]				50
Administrators, executives and technicians (inc. clerks) [a]	11.6		2,700	
of which:				
Central government employees [d]				300 (210) [e]
Other government bodies employees [d]				425 (300)
Manufacturing industry employees [f]				160 (88)
Banks and insurance employees [g]				75 (53)
Total			5,970	
Educated unemployed (below 500,000) [h]				

Note:

Estimated number of clerks in parentheses.

SOURCES:

[a] N.S.S. No. 14, pp. 46–55, 67–72.

[b] S. Natarajan, *A History of the Press in India* (Bombay: Asia Publishing House, 1962), pp. 306–307; GOI Ministry of Labour and Employment, "Report of the Working Journalists Wage Committee," (New Delhi: 1959), pp. 14–15. Estimated by assuming an average of 10 journalists for each of the 7000–8000 papers in India, and rounding.

[c] Based upon the number of law students and graduates in 1939/1940 and 1949/1950. I roughly estimate this figure. See B. B. Misra, *The Indian Middle Classes* (London: Oxford University Press, 1961).

[d] J. M. Healey, "Public Employment in India, 1951–60," *The Economic Weekly*, XV, No. 23 (June 8, 1963), pp. 925–926.

[e] Clerks in central government offices were about 70–75 per cent of the total white-collar employees, and I applied this proportion to other government bodies, and to banks and insurance companies; in 1956 clerks were about 53 per cent of the white-collar force in manufacturing.

[f] GOI Planning Commission, *Occupational Pattern in Manufacturing Industries*, 1956. In that year 8 per cent of the registered factory workers were in such positions. I apply this proportion of the 2.9 million registered factory workers in 1951; it is estimated that 70 per cent of these white-collar factory workers lived in urban areas, based on the all-India employment figures for manufacturing workers.

[g] Insurance figures: GOI, *Central Statistical Office Monthly Abstract of Statistics*, XVI, No. 3 (March, 1963), pp. v and vi: reduced and rounded downward to 40,000 for an earlier year. Bank Figures: "Award of Bank Disputes Tribunal," 1962, pp. 56–57: reduced and rounded to 35,000 for the earlier year.

[h] See Appendix C.

taxes as individuals or Hindu undivided families. Of these approximately 25,000 were in the highest bracket of over Rs 25,000 income per year. If the middle income group is defined as from those earning in the lowest annual income tax bracket of Rs 3,600 (or $720) to those in the bracket of Rs 12,500, this group totalled approximately 300,000 assessees. Of the total income tax demand in 1952/1953 from non-company sources, approximately 80 per cent was from assessees in the brackets above Rs 12,500.[27] The lower middle class, including most of the clerks and similar occupations, numbered approximately 6.0 million and earned below Rs 3,600 per year.

If these figures are even reasonably correct—and there is a great deal of underreporting and evasion of income tax in India—the picture they give of the income distribution within the urban middle class as of 1952 is a highly skewed one; the middle class consisted of a very small upper income group, a somewhat larger middle income group, and a great mass of low-paid, low-level white-collar employees, most of whom were employed by government and large-scale businesses. This very approximate statistical picture agrees with the impressions any observer would have in the large Indian cities.

Within this class it is possible to identify some of the more influential groups and to describe their attitudes as of 1950. In the next chapter I shall discuss the position of the Indian government officials under the British; following them in importance, especially in the light of the development programs, were the owners and executives of the large industrial firms. These were frequently centered about particular family groups and organized in the managing agency form. The data with respect to industrial ownership and control is admittedly approximate, but it indicates certain orders of magnitude with respect to the private industrial sector and the roles of various groups in the control of this sector. The share capital of all nongovernment-owned companies, both public and private limited, in 1951/1952 has been estimated at Rs 13,960 million. In that year the four largest business groups—Tatas (a Parsee firm), Birlas and Dalmia-Jain (Marwari groups), and Martin Burn (a Bengali-English group)—had full control by sole ownership or majority stock holding of approximately 15 per cent of the total share capital of both nongovernment public and private limited companies. They controlled 18 per cent of the total share capital of the public limited companies and approximately 17 per cent of the gross capital stock of these companies. If we

[27] This does not include tax on income earned from agricultural land, which is one reason for the smallness of the total. It would thus be largely urban income from salaries, dividends and interest, and urban rents. GOI, "Report of the Taxation Enquiry Commission, 1953/1954," II, chap. 9, Table 1, p. 135.

add the companies in which these same four groups had a minority interest, their proportionate share of the totals rises by about another 4 per cent. In that same year the leading thirteen business groups, including two English ones, had complete control of approximately 23 per cent of the total share capital of all nongovernment companies and had an interest in about 28 per cent; their proportionate share of the gross capital stock was probably similar.[28]

Although many of these leading business groups are from particular castes, communities (such as Banias, Jains, Parsees, Chettiars), and areas, and although they are influenced by such considerations, they have also been influenced strongly by factors beyond the ethnic; it is striking from what a variety of castes, communities, and areas they do come from. It is true that communities or castes frequently confined their investments in the prewar period to their own geographical areas—as the Gujerati businessmen from Ahmedabad tended to confine theirs to Bombay and Ahmedabad (but also overseas); the Chettiars of Madras, to Madras (but also overseas); the Naidus, originally of Hyderabad and then Coimbatore, to the latter city. The largest Parsee investment, the Tata Steel Plant, was in Bihar, however; and other Tata investments were in Nagpur, Ahmedabad, and near Poona, although the Parsees and Tatas lived largely in Bombay. The Associated Cement Companies—a combination effort of English, Parsee, and Gujerati capital—operated cement plants throughout India. Marwari businessmen originating in Rajasthan operated businesses and industries all over India. It is claimed that Marwari traders and moneylenders are in many villages of India; certainly the large Marwari entrepreneur families—Birlas, Dalmias, Jains, Singhanias, Bangurs, and others—have invested all over India in a wide variety of enterprises. These business groups are most important in the private sector, but their prestige has been low.[29] In part the relatively low prestige reflects caste traditions and the lower status of business activities, but in equal part it reflects a popular belief that business is crooked. This is supported by such facts as the very active black market during the war. Subsequently, distrust of business has been fed by charges of corruption, of hidden transfers of funds between companies within a group of evading income tax, of being speculatively rather than industrially minded—in the sense of being interested in short-term stock market investments and capital

[28] These data are from R. K. Hazari, *The Structure of the Corporate Private Sector* (mimeographed, 1963), chap. iv. The terms "public and private limited" do not refer to government or private ownership, but to company organization forms.

[29] Many of the following criticisms are usually made of groups other than the English agents and the Tatas, both of which have high prestige; they are made especially of the Marwari businessmen.

gains rather than longer-term industrial investments. The latter criticisms are largely alleged, they are rarely proved, although certain recent investigations—as of the Dalmia-Jain groups and of the postwar Mundhra enterprises—lend some weight to them in terms of past activities. But these criticisms should be placed in some perspective. In my opinion the activities of Indian business do not approach those of American businessmen in the American "robber baron" age; either many of the charges have never been proved and are based on unfounded gossip, or the activities mentioned were not illegal when carried out. Oftentimes the activities made a great deal of economic sense, for example stressing short-term gains when capital is scarce. A further factor contributing to the criticism is that these business groups are frequently national, and thus they meet the opposition of narrower regional or linguistic communities, which may be hurt by their activities.

As opposed to these criticisms, the great financial support that many of the business groups—especially the Birlas and the Gujerati businessmen from Ahmedabad and Bombay—gave to Gandhi and the Congress party in the period before Independence strengthened their prestige. Even more important, it increased their future influence with the Congress, not so much in terms of general policies, but in terms of access to leadership and administrative influence. It is important also, in describing business attitudes, to emphasize that although businessmen may be looked down on in India, there has always been a close connection between business and government with a tradition of monopoly and of government grant and control or both. The large business groups and the government both have felt more familiar with this system than another one—even apart from the gains therefrom or any ideological bias. Finally, leading business groups and others had prepared programs during the war and before Independence aimed at planned economic growth and industrial development after the war. So did the government itself during the war. These business-prepared programs, especially the Bombay Fifteen Year Plan of 1944, contributed to the acceptability of planning in India and the formulation of the Indian Five Year Plans.[30] They have also vigorously expanded their own industrial investments, especially in the postwar period.

The "intellectual community" is another very important urban middle-class group, transcending caste and narrow communal limits although often not the language limits. Since the early nineteenth century the large cities have seen the development of major intellectual movements, which

[30] On the Bombay Plan and its importance, see the recent article by one of India's foremost present planners, Tarlok Singh, "The Bombay Plan Recalled," *Eastern Economist,* XL, No. 22 (June 7, 1963), pp. 1176–1180.

were based either on western ideologies or on a reinterpretation of Hinduism to achieve some reconciliation of the Hindu classics and beliefs with the social and political ideas from the West. Calcutta, Bombay, Poona, and Madras were centers of such activities and of several extremely influential societies and religious groups. These were dedicated to social reform, to changing or breaking down the caste system, to wider education and greater opportunities for Indians under the British, and eventually to Independence. It is not mere coincidence that three nationalist leaders —Tilak, Gandhi, and Rajagopalachari—all wrote new translations or interpretations of the Hindu holy books, especially the Bhagavad Gita, stressing the activist character of Hinduism. Nehru's series of secular autobiographic and historical books of the 1930's and early 1940's had a profound influence upon the urban middle class as a whole, and especially the intellectuals. These national, intellectual figures were at first largely of Brahmin origin since this caste had the traditions and resources to take advantage of western education. However, other castes and communities also contributed: Gandhi was not a Brahmin, and the man most responsible for the Indian constitution, Dr. Ambedkar, was an untouchable.

By the time the Second World War began, India already had two Nobel Prize winners in literature and science, and a host of scientists and social scientists had returned from Western universities to train young men and women in Indian universities and research institutes including new ones they themselves set up. Simultaneously they had begun to carry on the research to provide a basis for the desired scientific and social reforms and to achieve a "New India." Shils estimates that there were about 700,000 university graduates living in India in about 1955, of whom approximately 60,000 were "professional intellectuals" and another 100,000 "intellectual consumers." These figures would have been somewhat lower in the period 1947–1950. He defines these groups as follows:

> including in that [60,000] figure college and university teachers, research workers in government and on the staffs of scholarly and scientific research institutes, applied scientists in industry, writers and analysts in the intelligence and publicity services, journalists, literary men, critics, scenario writers, painters and sculptors [plus] productive intellectuals on the bench and at the bar, in medical practice, civil service, in business and in active professional politics. . . . The consuming intellectuals whose vocations are not in themselves intellectual but whose training and disposition, whether connected with their occupations or not, lead them to interest themselves in intellectual matters must run into the neighborhood of 100,000. . . . In absolute terms, this is a large number, large enough to provide a modest audience for books and reviews, an audible voice in public opinion and above all for diversified and

stimulating intercourse. . . . [The] number is far larger than the intellectual class of any other of the underdeveloped countries of Asia and Africa.[31]

This group was largely educated in the interwar period, either in universities with many English faculty members modeled on English universities or in England itself. It was a group whose *lingua franca* was English and whose reading and ideas were often from English sources—frequently of the Labour party of the interwar period, since the members and periodicals of that party were most sympathetic to India. However, many also knew well their regional language and culture and strongly supported the renaissance of a regional literature and art. It was a group whose response to the west was often a mixture of great respect for western achievements, impatience with India's lag, and resentment of western rule and its ignorance of Indian achievements; and the resolution of this conflict varied among different individuals from the extremes of complete westernization to an abandonment of the west, but more often there was an attempt to reconcile on some level the western and Hindu ideas. Finally, it was a group many of whose members felt that India was capable of determining its own future, that its goals were not the same as England's, that England's rule was now a hindrance rather than a help in achieving those goals, and that with proper policies or independence India could once more become a great and powerful country. Also, with positive and forceful policies—to apply modern science, to develop industry, to introduce some form of land reform and coöperative land working under some vaguely defined socialist system—many felt that India was capable of increasing the economic well-being of its hundreds of millions of people to some level approaching that of the west: here Russia's example was important. Finally, there was a belief that these goals and hopes could be achieved by democracy in India, though there were also strong groups favoring more forceful and dictatorial methods.

The two middle-class groups described above were parts of the small upper and middle income class groups. The great mass of white-collar workers, who formed the bulk of the bureaucracy and the clerical and teaching staffs, were in the lower income groups. These struggled either to maintain a social position that their incomes made it increasingly difficult to support, or to reach a social position that their incomes make it difficult to attain. They were largely upper caste and therefore had a

[31] Edward Shils, *The Intellectual Between Tradition and Modernity: The Indian Situation* (The Hague: Mouton, 1961), pp. 20–21; on the relationships among the Indian intellectuals, new ideas of Hinduism, and modern India, for a summary picture, see K. M. Pannikkar, *The Foundations of New India* (London: George Allen & Unwin, Ltd., 1963), chap. 2.

belief in their own prestige and position; they were educated to some extent and literate, often averse to manual labor because of its low status, and, when first employed, relatively highly paid. But all these advantages were under attack after the war. A perceptive observer of the Bengali middle-class society of Calcutta writing in 1957 described the change in its position since the war.

> [The] contours of Bengali society were more or less clear and well-defined till [1939]. . . . [The] speed of vertical mobility in the urban society of Calcutta was much less in pre-1939 days than it was in the war years and than it still is in the post-war years. . . . The economic competition is becoming harder day by day for Bengalis in Bengal and at the hub of it, in the city of Calcutta, the Bengali middle-classes are being pushed through the borderline of gentility towards the working-classes. . . . [As] the frontier of gentility is changing and its scale is ascending, the struggle for climbing up to it and for preserving the status quo as well is becoming keener day by day. The crisis of Bengali gentility is therefore not only economic. It is a social and psychological crisis also.

The writer also referred to a study of budgets of white-collar central government employees in India's four major cities in 1946, which showed that at least 30 per cent of the families were in debt in other cities, but the proportion in Calcutta was 76 per cent. He noted that this "is also one of the most important reasons why 'trade unionism' developed among middle class employees all over India in wartime and [thereafter] and why Calcutta became one of its storm centers. . . . [But] it is very difficult for the Bengali middle-classes to . . . merge themselves with the working classes. They are therefore found to swing from right to centre and to left in the political field at surprisingly short intervals." The author concluded that these wide shifts indicated both the absence of any agreed-upon ideology among the Bengali lower middle class and its willingness to try any extreme to retain or improve its economic and social positions.[32]

This description of the Calcutta middle class and its attitudes is peculiar to Calcutta only in the degree of middle-class problems there, not in the type of problem. The members of the lower middle class throughout India were deeply involved as leaders and members in anti-Brahmin movements in the south and in Maharashtra, in militant communal movements in North India near New Delhi, and in extreme left-wing movements in West Bengal and Kerala.

Finally an important group in this lower middle class was the educated unemployed. By 1939 there were 385 colleges in India with 145,000 students including 23,000 graduate students; there were 15 million students

---

[32] B. Ghose, "The Crisis of Bengali Gentility in Calcutta," *The Economic Weekly*, IX, Nos. 26, 27, and 28 (July 6, 1957), pp. 821 ff.; quotes are from pp. 823 and 826.

in primary and secondary schools. The training was largely in the liberal arts. Of the 17,000 professional students in 1939/1940, close to half were studying law; a third, medicine; and about 10 per cent each, teaching and engineering. Although this education system trained many of the leaders of India, it was very difficult for many of the graduates to get any jobs. English firms prior to Independence were often reluctant to hire Indian technical people; graduates in law and liberal arts entered an already overcrowded market. In 1955 there were 200,000 registered educated unemployed (defined as those unemployed with at least a high school education or its equivalent); at Independence it must have numbered at least 100,000. The British government had already begun to worry about the educated unemployed at the time of the First World War, and this worry increased. It was this group that provided the members for the political terrorist and revolutionary movements in Calcutta and other cities in the interwar years. Although nationalism provided an outlet for the members of this group, with Independence the Indian government had to face the problems they posed.[33]

THE URBAN WORKING CLASS

Like the middle class, the urban working class will also be defined by the groups it includes. It includes operating workers in factories and handicraft industries, construction workers, such relatively unskilled service workers as messengers (peons), cleaners, and scavengers, as well as more skilled washermen, barbers, and cooks, peddlers in the commercial field, and the totally unskilled workers. As defined here it excludes people in agricultural occupations. The National Sample Survey makes possible a rough quantitative estimate of the numbers in the various occupations in this class in 1953/1954 by providing percentages, which may be applied to the total urban gainfully employed labor force figure of 24 million in that year. This estimate is presented in Table 7.

With respect to the incomes of these workers relatively little is known, but it would unquestionably be correct to regard an income of Rs 200 ($40) per month as the upper bound for all these workers in 1953/1954, with most worker incomes being below Rs 100 per month.

Little is known as to what extent these workers function on a caste basis, a communal basis, or a class basis. It is possible that one of the incentives to urban migration, especially by the growing number of higher

---

[33] B. B. Misra, *The Indian Middle Classes* (Bombay: Oxford University Press, 1961), pp. 304–305; K. Davis, *The Population of India and Pakistan* (Princeton: Princeton University Press, 1951), chap. 17; GOI Planning Commission, *Outline Report of the Study Group on Educated Unemployed,* 1956, chap. 3; N. K. Bose, "Social and Cultural Life in Calcutta," *Geographical Review of India* (December, 1958), pp. 24–26.

TABLE 7
BREAKDOWN OF URBAN WORKING CLASS OCCUPATIONS, 1953–1954

Occupational Group	Per Cent of Urban Labor Force	Number of Workers (thousands)
Operatives and artisans		
(including supervisors) [a]	15.2	3,650 (1,800) [b]
Household industry workers [c]		
Food products	3.5	840
Textile products	5.4	1,300
Service workers	13.4	3,220
Hawkers, peddlers, etc.	2.4	580
Building industry workers	1.7	410
Unskilled labor	8.1	1,940
Total		11,940
Unemployed		about 500,000+

Notes:

[a] Operatives and artisans: "Workers engaged in operation of power equipments or manually operated plant and machinery (excluding supervisory work), smiths, drivers, boatmen, technical apprentices, artisans (excluding washerman, barber, cook and weaver), craftsmen, sawyer, workers engaged in general engineering work." Supervisors are probably foremen.

[b] Number of employees in registered factories are in parentheses.

[c] Manufacturers of food products: "Husker and parcher of food grains and pulses; crusher of oil seeds, confectioner, gum and candy makers, baker, producer of indigeneous liquor, and makers of . . . tobacco products."

Manufacturer of textiles: "Ginner and cleaner of fibre, spinner, weaver, dyer, and printer, carder and reeler."

From the above definitions I would assume that the "operatives," etc., include factory workers; while the "manufacturers" include handicraft workers in the food and textile industries. Overlapping may exist; some of the manufacturers may also be employers.

SOURCE:

N.S.S. No. 14, pp. 46–55, 67–72; see also Appendix D.

caste non-Brahmin Hindus, is the hope of maintaining or improving village caste position in the economic area. Related to this is the admittedly very crude estimate that the flow of remittances from urban workers, often single males, to their home villages, was on the order of Rs 1,200 million in 1952/1953. Within the city, however, communal factors or very broad caste differences, between Brahmins and others, are probably more important than village caste factors in influencing urban worker behavior—except probably in the case of marriage.[34]

To what extent have the workers functioned as a class through such worker organizations as unions? The statistics on Indian union membership are very poor, making general conclusions difficult. Unquestionably within the working class, as distinguished from the lower middle-class

[34] M. D. Morris, "Caste and Evolution of the Industrial Work Force in India," *Proceedings of the American Philosophical Society*, CIV (April, 1960), pp. 124–133, presents the most thorough discussion of the role of caste in urban labor relations. K. N. Raj has made this rough unpublished estimate of remittance flow.

clerks, the factory workers would be the most organizable into unions, since others are too dispersed or too dependent on the employers. In 1950 there were about 1.8 million workers in registered factories earning less than Rs 200 per month.[35] These were probably all in working-class occupations but may have included some clerks.

Although trade unions have existed in India for some time and, at the end of the war, had a claimed membership of close to one million workers, in fact little information is available on their internal operation. It is known that unions have been, and are, closely related to Indian political parties. Gandhi was the leader in the famed Ahmedabad strike that led to the formation of the Textile Labor Association; Congress party leaders like Nehru, Patel, and S. Bose played an important role in founding and leading national trade unions at an early date. Communist and Socialist leaders, whether in or out of the Congress party, were major influences, and the Communists controlled the All-India Trade Union Congress at the end of the war; this forced the Congress and Socialist leaders to set up their own unions after 1947, and in some areas this was the initial Congress entry into unions.

The leadership of unions was and still is rarely of working-class origin. B. B. Misra states that:

> The growth of industries . . . brought into being a considerable number of salaried employees [of the lower middle classes] below the managerial and supervisory levels. . . . There came into existence a hierarchy of labour leaders whose interest it was to forward the claims of their own class by an alliance with industrial and other workers. This alliance necessitated the organization of trade unions and factory legislation. Industrial strikes became an additional weapon designed not only to wrest economic concessions but also to threaten the continuance of foreign rule.[36]

Weiner points out that not only unions but also other pressure groups—such as peasant organizations, refugee groups, and the like—are often controlled by "full-time professional politicians, frequently with private incomes from the land, who are college or at least secondary school graduates. Generally, by reason of their income, education, family background, landownership, or caste, they are of higher social status than the groups that they lead." Weiner explains this largely on grounds of status considerations: "the unwillingness and inability of people at the bottom of the social ladder to communicate directly with those of higher caste, economic standing, and power—make it difficult for factory workers [and

---

[35] GOI Labour Bureau, Ministry of Labour and Employment, *Digest of Indian Labour Statistics* (New Delhi: 1961).

[36] B. B. Misra, *The Indian Middle Classes*, p. 397.

other low status groups] to deal directly with management or government."[37] Workers therefore need an intermediary group of educated leaders of relatively high status who can talk to industrial and government figures on an equal footing. It is necessary however that these leaders on a regional or local level know the regional languages of the worker, as well as the higher authority's language—most likely English—and this would inevitably mean that regional or local union leaders are of the same community as the workers. But the national union leaders are also generally national party leaders, and as such they transcend narrow community lines.

Thus, although the unions function as a class, the interaction between leaders and members frequently crosses class lines, and ideological or communal relationships may be more important. However, in some areas unions have apparently split on caste lines. The union members themselves are interested in such bread-and-butter issues as wages, bonuses, and security; ideologically, however, the unions reflect their leadership, and their affiliated parties. But where there has been a sharp conflict between the bread-and-butter desires of the workers and the union leaders' positions as members of the parties, the workers have changed unions. Furthermore, in voting since Independence, workers from rural backgrounds often support a party like the Congress, which is not based on class, rather than an out-and-out class party.[38] Although there was a working class in urban India, its role and influence as an organized group with its own interests was relatively slight and quite diffused.

---

[37] Myron Weiner, *The Politics of Scarcity, Public Pressure and Political Response in India* (Chicago: University of Chicago Press, 1962), pp. 90–92. Quotations are on pp. 91 and 92.

[38] For the prewar history of unions, see Charles A. Myers, *Labor Problems in the Industrialization of India* (Cambridge, Massachusetts: Harvard University Press, 1958), chap. 4; M. Weiner, *The Politics of Scarcity,* chap. 4; and Gene D. Overstreet and Marshall Windmiller, *Communism in India* (Bombay: The Perennial Press, 1960), pp. 367–374.

3

British Government and the Congress Movement

BRITISH GOVERNMENT

Prior to Independence the British were the governors of India. In 1931 there were approximately 115,000 British Europeans in India, about 70,000 of whom were in the army and less than 50,000 in civil government and unofficial pursuits.[1] With the war and the expanded activities of the British this number probably rose during the forties.

The head of the British government in India was the Viceroy, the direct representative of the British crown and less directly of the British cabinet and Parliament. Without entering into complexities, the Viceroy and his Council had executive power in the central government; and under the 1935 Government of India Act,[2] he, with the advice of a Legislative Assembly, had responsibility for the large variety of decisions and functions in the power of the central government, especially foreign affairs, defense, and so on. He also appointed governors for the various provinces. These governors had various residual powers; but the provinces were governed by popularly elected ministries, and these had power in a large number of local fields such as agriculture. Provincial legislatures were elected by a limited popular suffrage of 35–40 million voters throughout India, and the majority party or group in these legislatures chose the provincial chief ministers and cabinets. The Congress, functioning as a party in the 1937 provincial elections, was successful and set up provincial ministries. Apart from these more-or-less elected organs of government, there were the 600 scattered princely states covering over 500,000 square miles and inhabited by almost 100 million people. These were outside the 1935 system, for they were governed directly by their princely rulers under treaty arrangements with the British, by which the British government kept direct control

[1] K. Davis, *The Population of India and Pakistan,* pp. 95–96. This latter figure includes family members, so the number of male civilians would be on the order of 15,000 to 25,000 only.

[2] Much has been written on the provisions of this act. I have used as convenient summaries the relevant chapters or pages in V. P. Menon's two books, *The Story of the Integration of the Indian States* (Madras: Orient Longmans, 1961); and *The Transfer of Power in India* (Calcutta: Orient Longmans, 1957); and Hugh Tinker, *India and Pakistan* (New York: Frederick A. Praeger, 1962). A more recent book on the British relationship to India is M. and T. Zinkin, *Britain and India* (London: Chatto and Windus, 1964).

over their foreign affairs and defense. Since these provided for a British representative who could recommend their replacement, the British in fact had ultimate power in these princely states; and the princes ruled, but subject to these restraints. The quality of the rule varied from state to state, with a few very well run and others poorly run; some were one-man despotisms, and others ruled through ministers: in all, the rule was comparatively personal and less bureaucratic than in British India or subsequently free India.

This was obviously a very complex system. It was hopefully a transition toward greater self-government. In fact, the Second World War brought it to an end. The one-sided involvement of India in the war by the Viceroy was legal but unpolitic; the subsequent resignations of the Congress ministries and the initiation by the Congress of the "Quit India" movements to force the British out of India led to its breakdown. Nevertheless, as embodied in the 1935 Government of India Act, this system provided the framework for the Indian Constitution of 1950; furthermore the experience gained by the Congress leaders in running provincial governments proved invaluable on their subsequent assumption of national power after Independence.

However, the main contribution of the British to the government of an independent India was not a constitution; rather it was the handing over of a trained and very able bureaucracy and a set of policies. Under the British, this bureaucracy was largely independent in purely administrative matters of internal Indian group conflicts and politics,[3] and thus it remained relatively nonpolitical. Philip Woodruff, himself a former Indian administrator, has entitled the second volume of his history of the Indian Civil Service (ICS), which was the core of the British administration, *The Guardians*. Two comparable names still heard in India to describe the ICS are "the steel frame" and "the heaven born." All imply its incorruptibility, its freedom from politics and regional or communal interests, and its efficiency.

In 1947, upon the eve of Independence, there were 1,064 members of the ICS, of whom approximately 425 to 450 were Indians. Of those Indian members, approximately 300 became members of the Civil Service between 1929 and 1939. The ICS members who chose to work in India rather than Pakistan after 1947, including a few British who did not choose to retire, formed the nucleus of the independent Indian administrative bureaucracy. In 1947, the employees in the central government alone —excluding the rail, post and telegraph, and defense departments—num-

[3] In larger matters the British introduced the practice of "reserved" seats for minority groups, and have been charged with playing off one group against another —Moslems against Hindus, princes against Congress—to maintain their rule.

bered 157,000. Of the ICS who opted for India, many have undoubtedly retired or died, but a rather rapid survey of the *Times of India Yearbook* for 1961/1962 found from 145 to 150 ICS men in key roles in the central and state governments as of that date.[4] These include positions of Secretary, Additional Secretary, and Joint Secretary in the central government, and Chief Secretary, Department Secretary, and department heads in the state governments. This is at present the most important group within the Indian bureaucracy. Assuming that most of the members were appointed between 1935 and 1940 and at about the age of 20–25, one can assume they will continue to be important for the next ten years. It will therefore be of some value briefly to indicate their general background and their approach to Indian problems.

First, these men were selected after a very difficult competitive examination, in competition not only with the other Indians, but with the ablest English graduates, since the ICS was one of the top services of the entire imperial administration. Choice was not made on a personal, family, or community basis. Second, these men were not specialists: they were specifically chosen as generalists, in the British tradition. Many of the members were originally trained as classicists, philosophers, and mathematicians, as well as in some more specialized areas. Third, the Indians chosen would obviously be western educated, either in Great Britain or India; after selection they received additional training at Oxford or Cambridge. They thus were probably from economically upper-class families of high caste, but as a group they were probably among the freest in India from traditional caste and community limitations.[5]

One English ex-officer of the ICS has written a most interesting memoir giving his impressions of the Indian ICS officers whom he knew from 1938 to 1947.

> With the exception of [one man], none of those whom I knew well were the sons of great houses. Their fathers were lawyers or officials or occasionally businessmen. . . . Sometimes my contemporaries' parents were old-fashioned, but more often the break with the past had come in their parents' generation. But on the whole they were no longer bound by the traditional ritual. They

[4] This is probably an understatement since it does not include foreign service officers; and some of the listed officers may not have placed ICS after their names.

[5] The above paragraphs are derived from Philip Woodruff, *The Guardians,* (New York: St. Martins Press, 1954), especially pp. 348–366; V. P. Menon, *Indian Administration, Past and Present,* Forum of Free Enterprise pamphlet (Bombay: 1958); A. D. Gorwala, *The Role of the Administrator,* Gokhale Institute of Politics and Economics (Poona: 1952); R. Braibanti, "Reflections on Bureaucratic Reform in India," and S. P. Jagota, "Training of Public Servants in India," both in *Administration and Economic Development in India,* eds., Braibanti and Spengler (Durham: Duke University Press, 1963), especially pp. 1–74.

were normally fiercely against the old exclusiveness. They were Indians first and foremost. . . . Untouchability was a matter of shame, the restrictions of caste something to be swept away. Communal divisions, they thought, were purely a British creation.

[The] tradition [of the Service] emphasized the District and not the Secretariat. . . . To want to go into Bombay was thought soft; any hint of the desire to go to the Government of India [in New Delhi] was sure proof of unsoundness. The place of a good officer was on his horse, out in the villages, listening to the grievances, remedying injustices, settling disputes, getting to know how the people felt and what they wanted.

The belief that only the peasant really mattered was perhaps romantic. . . . India could not be governed in forgetfulness of the lawyers, the moneylenders and shopkeepers whom the average British official affected to despise. Indeed the official's outlook was in some ways rather close to that of the Gandhian; that is why he nearly always felt a sympathy for the true Gandhian he most certainly did not feel for the ordinary run of Congressman. The ideal of the Official . . . was so often the self-sufficient village, with its patriarchal and self-respecting leaders . . . [but] he was [never] able to see the economic point of hand-spinning. . . . Tariffs should be raised and India industrialised as quickly as possible. Women should be treated as equals . . . and money-lenders curbed. Many were socialists and planners. Some were disturbed about the difference between their standard of life and those of the people. . . . But the whole range of what Mr. Nehru wanted has been exactly what the I.C.S. has always wanted: State capitalism, the destruction of caste, the reform of Hindu law, a proper place for India in the world, industrialisation, land reforms, universal education, fertilisers, and cooperative credit and bunding—these were dreams of our youth. . . . My contemporaries were an elite such as no other country in the underdeveloped world has had. But there were very few of them. . . . [That was] changed by the war. . . . [The] Government [during the war] gave itself over to what was in effect one of the grandest training programmes in history. In 1939, had the British gone over night, India would have found itself desperately, perhaps fatally short of all the people needed to run a modern State. By 1946, there were enough. . . . From then on, British rule was dead, for the British no longer had a function to perform.[6]

The ICS was an administrative elite: it came from a similar Indian social base, it was responsible to an outside government, it was chosen as competitively superior and believed in its own superiority, it was trained to a peak in the function of decision-making; its members, working closely together and knowing each other well, constituted almost a new group of

[6] M. Zinkin, "Some Aspects of Change in Indian Society, 1938–60," *Changing India,* ed. N. V. Sovani and V. M. Dandekar (Bombay: Asia Publishing House, 1961).

Brahmins. In many cases its attitudes were reinforced by the traditions of Brahminism. Prior to the war the primary function of the State and the administrator was:

> . . . the maintenance of order and the enforcement of the due process of the law. . . . The second main task . . . was the maintenance of efficiency in all basic matters. . . . A special characteristic of this period was the great emphasis laid upon the necessity for complete impartiality as between man and man, community and community. The ideal of justice was very firmly held aloft. . . . The alteration of fixed economic patterns was no part of his task. . . . [The] tendency was to leave economic matters, whether in the urban or rural centers, severely alone.[7]

In pursuance of the first task, the British set up their own legal system, which was staffed initially by English justices on the higher level and introduced into the districts by the district officers. As pointed out earlier, the introduction of English law and the withdrawal of legal support for the caste system created an alternative system to caste and, thus, introduced a fundamental contradiction to the working of the caste system.

Even prior to World War II there were exceptions to the purely "nightwatchman" role of the government toward the economic system. Several British officials sought valiantly but without success to introduce cooperation in their districts—for example, by taking steps to curb moneylenders. The Second World War led to a far greater and more direct government involvement in the economy. The production, distribution, and pricing of commodities such as food, grains, textiles, steel, and coal were controlled; the issue of capital and foreign trade were closely directed and supervised; and institutions were set up to perform these control duties. These control systems were in many cases drafted and enforced by Indian administrative personnel, and this experience provided the framework and, in many cases, the methods for the controls of the postwar and planning periods. The tendency to favor controls was supported by an administrative distrust of business—in part derived from Indian Brahminism, in part from British upper-class attitudes.

Well before the war an extensive railroad system had been constructed by the British, and this system had been largely nationalized and run by the government; extensive systems of post offices and of telegraphic, telephonic, and radio communication were being operated by the government; and a road network had been at least partially constructed. Although the initial purposes of some of these systems may have been political, they yielded major economic benefits. During the period of British rule the top positions in these government-owned operations were largely held by

[7] A. D. Gorwala, *The Role of the Administrator,* pp. 4–6.

English administrators and technicians; but many Indians were trained in running these services, and they succeeded the British when the latter left. The form of the administrative mechanism for running these operations was largely maintained after Independence.

Although I have laid greatest stress upon the administrative personnel left by the British, the British also trained the present leaders of the police force—there was a select Indian Police Service comparable to the ICS—and the officers of the Indian Army. Many of these latter were educated in British training academies, and served with the British army during the Second War; and like the ICS officers, they formed a close group that was quite westernized, and—certainly prior to Independence—both removed from civil politics and with an attitude that they should not be involved. Their influence on economic development policy at present is, of course, small. Their political role in the past has been small in a positive sense, but they contribute necessarily and directly to maintaing the internal security of the country. Thus the British left India with a framework of government and law, the nucleus of a very skilled bureaucracy and security force, and a set of government functions and bureaucratic attitudes that have profoundly influenced actions and policies since Independence.

Although Indian support for British departure in 1947 was general, many groups opposed Independence. Foremost of these groups were the princes, defined to include not only the rulers, but also many of their bureaucracy and many of the peasants in the princely states. Another rural group fearful of Independence were many of the very large landowners, especially the large zamindars who performed a tax-collecting function under the British; in the urban areas many of the wealthy upper-middle-class western-educated Indian lawyers and professional men played a prominent political and social role in British India and regretted the split. Many of these combined large landholdings in rural areas, from which they received *rentier* incomes, with high professional and social positions in the cities. Obviously the British business group, which played such a large role in prewar India industry and banking and had strongly influenced the British government's economic policies, feared their investments might be jeopardized with Independence. All of these felt they would lose their existing social, political, and economic positions if the British rulers left. On the communal level, such groups as the Anglo-Indian community and some of the smaller communities felt that in a Hindu India, strongly influenced by communal and caste pressures, they would lose their positions and possibly even some of their privileges. Even the untouchables felt threatened by a Hindu government, and Gandhi developed his movement to break down untouchability in part to convince them otherwise. The Moslems, the largest of the minority communities, were concentrated in several provinces in then Northwest and Northeast India, and they set up their own movement for a separate Moslem state once the Congress movement appeared to be achieving successes. Strong parts of all these groups felt threatened by the achievement of Independence, and some chose to depart after it.

THE CONGRESS MOVEMENT

By the time the British were ready to leave India, the Congress movement had become the major political representative of the non-Moslem Indian population. The Congress led the demand for Independence, it formed the government of India when the British departed, and it was converted into a democratically elected governing party after Independence.

Initially the Congress, founded in 1885, was not in favor of Independence. In fact, it was organized with British support, as a "controlled" vehicle for expressing protest. The leadership was "moderate" and upper middle class. Close to 40 per cent of the delegates to the annual sessions between 1892 and 1909 were lawyers, another 25 per cent were of other urban professions—journalists, doctors, and teachers predominating. Of the remainder about 25 per cent were landed gentry, and 15 per cent were from commerce. Among the leading financiers of the party were wealthy princes, large landholders, and the Tatas of Bombay. The leadership was largely western educated, and the aims of the party were an improvement in the position of the educated Indian upper class in the Indian administration—opening ICS to Indians was a major goal—and social reform. But their desires were within the framework of continued British rule in India. At the same time, the Congress at this early stage tended to oppose land reform or steps to curb the power of the zamindars, opposed legislation to improve the condition of factory labor, and favored a protective tariff for industry, abolition of excise taxes on cotton textiles, and other policies to encourage the development of Indian industry and entrepreneurship. Thus, the interests of the educated Indian for an improvement in his opportunities and the interests of local entrepreneurs in the development of industry coalesced. The early Congress party, though composed largely of professional classes, was

> . . . a superior ally of the Indian bourgeoisie. . . . The holding of an industrial conference simultaneously with the annual session of Congress in 1905 and the launching of the boycott of British goods in 1906 were by far

the most significant expressions of the alliance between the intelligentsia and the bourgeoisie.[8]

But the moderate agitation was ineffectual even in terms of the limited goals. The moderates with their western education and upper-class background were too far removed from the Indian public to create a mass movement, and they had only a minor effect upon the British. The next group of participants and leaders in the Independence movement was from the lower middle class, which had not achieved status and position under the British. This included the high caste Bengalis and Maharashtrans, who were educated and urbanized but either were unable to find jobs or, more often, found lower income jobs. The lower middle class was also a Hindu group, more traditional than the previous moderate groups; it saw the Hindu values threatened by westernization, and sought a contemporary interpretation. From this group came both the leaders of a violent terrorist movement that centered in Calcutta and the leaders of a Congress movement that was far more active—one that supported violence. The major Congress leader of this period was Tilak, a Maharashtran Brahmin, who also wrote a highly influential activist reinterpretation of the Bhagavad Gita and revived several major Hindu holidays in Maharashtra, so that they had a mass following in the urban areas in Bombay. During this period, Independence became a goal of the Congress, and it thereby lost the bulk of its support from the princes, the larger zamindars, and many of the successful upper-class moderates. Bailey's description of this process in Orissa is of great interest:

> In the early years middle class political leaders of the coast received both the patronage and the active help of some important Raj families. . . . In the second decade of the twentieth century politics on the coast began to shape into the more definite form of the Independence Movement. This on the whole, neither the Rajas of the Feudatory States nor the Zamindars of the coastal plains . . . supported. . . . [Most] were active sympathizers of the British. . . . As the Independence Movement developed and became more forceful and aggressive, the Congress of the coastal districts . . . [sought] to subvert the authority of the Rajas [of the Feudatory States] who were regarded as the supporters of the British; . . . [and] to bring about social and agrarian reform and ultimately to establish some form of representative government in [those] states.[9]

---

[8] On this early stage, see B. B. Misra, *The Indian Middle Classes*, pp. 352–367; R. I. Crane, "The Leadership of the Congress Party," *Leadership and Political Institutions in India*, eds. Park and Tinker (Bombay: Oxford University Press, 1960). pp. 170–172.

[9] F. G. Bailey, "Politics in Orissa," Part III, *The Economic Weekly*, IX, No. 39 (September 19, 1957), p. 1297. This is the third part of a series of nine articles on Orissa politics by Bailey which appeared from August 29 to November 7 in that magazine.

The effect of the First World War was to moderate the independence movement in the support of England in the war, but the disappointment with the British policies on Indian home rule after the war stirred up increased tensions and greater violence, which was forcibly repressed. At this time, too, Tilak died, and Mahatma Gandhi took over the unquestioned leadership of the Congress movement. This had certain major effects. Gandhi introduced the philosophy of *satyagraha,* with its technique of nonviolent resistence, which the British had difficulty responding to. Gandhi himself was a non-Brahmin from Saurashtra; he was in fact a Vaishya, or *bania,* of the trading caste, and thus appealed to a wider group than the previous leadership. He also appealed to the influential Gujerati and Marwari businessmen and industrialists who were of similar caste and from nearby areas. Gandhi's orientation has been described in the following terms by a leading contemporary Maharashtran intellectual:

> [as] pronouncedly Hindu, anti-intellectual and more backward looking than that of the generation that had gone before. . . . It gave to some of the religious terms of Hinduism . . . a political meaning, and political doctrines were clothed in spiritual terms. Its intellectual content was very much inferior, but its emotional appeal was great. . . . It can perhaps be said that without a watering down of the severe intellectualism of the earlier movement, it could not have appealed to the masses. . . . The Indian National Congress gradually became a predominantly Hindu body. The Muslims drifted away.

While the movement became increasingly Hindu, Gandhi tried to rid it of

> . . . violence, racial hatred, and communal jealousy. . . . [He] endeavored to convert [the Congress] into an instrument intended not only to fight for political freedom but also to reconcile, through a religious approach, the conflicting interests of labour and capital, landlord and tenant, Hindu and Muslim, Brahman and *Harijan.*[10]

Gandhi made the Congress a movement rather than a narrow pressure group in the traditional sense. And under his leadership it was a movement for independence under which class and caste interests were, to a large extent, subsumed, even though they existed. As pointed out, there were certain classes whose support it did not have: the rulers of the princely states were almost entirely out of it, and likewise most of their peasant subjects who preferred traditional royal rule; in the urban middle class the large rentiers, western educated and successful, were also out of

---

[10] The first quote is from N. V. Sovani, "British Impact on India After 1850–57," *Journal of World History,* Vol. II, No. 1 (April, 1954). Sovani is a leading Maharashtran Brahmin intellectual. The second is from B. B. Misra, *The Indian Middle Classes,* p. 399. See also S. H. Rudolph, "The New Courage," *World Politics,* XVI, No. 1 (October, 1963), pp. 98 ff., on Gandhi's psychology.

it. They lived in the large cities and received income from landed estates. These were the moderates, who were neither comfortable in the Congress nor had the same interests as the lower middle class, and they wanted to maintain ties with the British and work within the system. The Indian members of the ICS and other government organizations, who did not choose to resign and honestly felt they could do the best for India within the government, could not join the Congress. Among the peasants, the great bulk of the landless laborers and poorest peasants knew nothing of the Congress and were indifferent to it.

Within the Congress, however, were representatives of many of the classes described in Chapter 3. Gandhi appealed to the landowning and operating peasants and to the middle-class group in the smaller towns descended from these peasants. Several of his leading satyagraha campaigns—the Champaran campaign in Bihar in favor of the plantation laborers against the English indigo planters, and the Bardoli campaign of 1928 in Gujerat to relieve the peasants of land taxes in a time of distress—arose from peasant grievances. These campaigns brought forth such leaders as Rajendra Prasad and Vallabhai Patel, both local middle-class lawyers of peasant origins. These leaders as well as others on a lower level were also attracted by the Gandhian social movement to revive the virtues of the peasantry, and by Gandhian economics with its emphasis upon spinning, home industry, and prohibition; Gandhi made the Congress a social reform movement that attracted the social worker and reformist. As part of the social reform movement Gandhi advocated, for political as well as social reasons, the improvement in the social condition of the untouchables—named by him the *Harijans* or "children of God"—by permitting their entry into Hindu temples; and similarly he advocated raising the social position of women and associating women with the Congress party. Together with a sister of the leading mill owner, he also led the Ahmedabad workers in a strike against the Ahmedabad mill owners and, in the process, created a new type of labor union in Ahmedabad. Finally, although creating a less secular movement, he sought to prevent it from becoming a purely Hindu movement by insisting on joint action with the Moslems and protection of Moslem rights. In many of these social reform aspects of his movement, Gandhi had the support of the westernized intellectuals and attracted a moral reform group into the Congress party—a group of which Vinoba Bhave was a leader.

At the other extreme within the Congress party were the Congress Socialists, which included at that period such men as J. P. Narayan, M. Masani, and R. Lohia. These were not joined formally by such leaders as Subhas Bose and J. Nehru, but together they provided a mostly western-educated, urban, and intellectual leadership that favored socialism, large-scale industrialization, and planning. There were sharp conflicts between this latter group and the group represented by V. Patel and R. Prasad, which normally had the majority: in 1939, Subhas Bose, who was largely supported by the more urban, more westernized faction, was forced out of the Congress Presidency and the party—even after apparent success in challenging Gandhi and Patel. The bridges between the two groups were Gandhi himself and Nehru, who never formally joined the socialists. This more socialist group also sought to organize the landless peasants and the farm laborers, as well as the factory workers, and in the late 1930's it worked closely with Communists who joined the Congress in large numbers as part of the United Front tactic. The Communists left the Congress over the war issue—after Russia was attacked, they supported the war—and won control thereby of much of the existing national trade union movement since alternative leaders were in prison or underground.

Another bridge between the Gandhians and the intellectual wing was, in a peculiar fashion, the large industrialists. The Gujerati, the Marwari businessmen attracted by Gandhi himself, and the Parsees favored a program of development of Indian large-scale industry by tariffs, subsidy, and similar positive policies; they were all inevitably in close contact with British industrial and economic leaders and resented the dependence. These were the financiers of the Congress party. Although they looked down on the intellectuals and the intellectuals in the party looked down on them, their advocated policies concerning industry tended to agree more with the intellectuals than with Gandhi's. Still, they were also in a close relationship, both personal and financial, with Gandhi and Sardar Patel.

These major class interests within the party were interwoven with communal, linguistic, and caste pressures and counterpressures. The largest minority communal group within India was the Moslems, and although the Hindus and the Moslems in the 1920's worked together in the Khilafat religious agitation, by the middle of the 1930's the agitation for a separate Pakistan had begun; and by the late 1930's the Moslem League had become a movement representing most Moslems. This was encouraged by the failure of the Congress party in the United Provinces, now Uttar Pradesh, to take Moslem League members into the provincial cabinet in 1937.

Among the Hindus themselves communal pressures played a great role within the Congress. In the South, the fact that initially the leadership of the Congress was Brahmin led to the formation of the Anti-Brahmin Justice party; and the Anti-Brahmins did not join the Congress in large numbers until shortly before 1937—in fact, not until the provincial elections when Congress looked as if it would win. In Maharashtra the initial leadership of the Brahmins in the Congress had discouraged non-Brahmin membership, and Gandhi's non-Brahmin leadership brought in the landowning Marathas. In Andhra, initially, landowning and cultivating Kammas and Reddis united against the Brahmin leadership; but after the Brahmins were ousted the Reddis captured control of the Congress, and the Kammas became the leaders of the Communist party in that area. In Bihar the leadership of the Congress was rural; although initially the Kayastha caste of scribes provided the leaders, other landholding castes—Brahmins and Rajputs—entered, and the struggles within the party reflected the struggles among those castes. In Bombay the Parsee leadership was initially strong, but the rejection by the Congress of a Parsee claimant for the chief ministership of that province in 1937 was claimed by the Parsees to be on communal grounds, and that alienated them. Within Orissa the conflict between Karans, a writer caste, and Brahmins contributed to splitting the Congress party in 1942, and the Brahmin leadership joined the Moderates to form a pro-war government.

But in spite of these splits and factions, under Gandhi the leadership avoided divisive policies, and there was unity, in general, in favor of independence. In fact, more limited groups that threatened this unity, such as the one led by S. Bose, were forced out. With respect to the peasants and land policy,

> . . . so long as peasant interests were adversely and directly affected by government [or by British landowners], the Congress . . . defended peasant interests with vigor. But where peasant interests were circumscribed by landed interests, the Congress under Gandhi counseled mutual trust and understanding.[11]

The Congress Socialist Group within the Congress and the peasant union, Kisan Sabha, which it supported, advocated drastic land reforms: in Bihar this group wanted the abolition of the zamindari system and elimination of agrarian debt, as well as more immediate reforms. However, the more conservative Congress ministry of 1937 in Bihar in effect acted as a neutral party. It both restricted Leftist agitation and negotiated an agreement between the wealthy zamindari and the Leftist groups to moderate the stringency of the reform legislation that was in fact passed in Bihar.

With respect to labor and capital, Gandhi developed the unifying theory of "trusteeship." Although he recognized the evidence of class struggle, Gandhi stressed that both capital and labor were necessary and that the laborers could strengthen their position, first by combining among themselves to act as a unified group and then by cooperating with capitalists to increase their joint wealth. For this joint wealth the capitalists were

[11] Walter Hauser, "The Indian National Congress and Land Policy in the Twentieth Century" (Mimeographed paper, prepared for the December, 1962, meetings of the American Historical Association), p. 4.

merely the trustees, according to Gandhi, and not the uncontrolled owners. Related to this was a policy of nonviolence in dealings between labor and management. The operation of the satyagraha philosophy in practice was shown in the Ahmedabad strike of 1918, where Gandhi got both sides to agree to a settlement that was, he felt, both justified and not inconsistent with either side's demands. In the process of the strike he founded the Ahmedabad Textile Labor Association, which carried out his ideas of united labor action. This trusteeship theory opposed Marxian "class struggle," and made it possible to win the support both of India's leading capitalists and strong factory-worker groups. Here again the Congress Socialist group and the Communists disagreed, and they formed "class" unions that carried out violent strikes against management. But the important political result was that in this area, too, Gandhi prevented the conflict between labor and capital from undermining the Congress.[12]

Thus in areas of agricultural policy and labor policy, the Congress under Gandhi developed both a theory and a style of compromise that made it possible to avoid conflicts that would seriously split the independence movement. Similarly, caste and community differences among the great mass of the Hindu population were not allowed to split the independence movement, even though the linguistic principle was accepted as the basis for the future reorganization of India. Although most Moslems left the Congress and formed an independent party, the Congress always claimed to be a secular party and such Moslems as Maulana Azad held very high Congress positions.

There was at least some minimal agreement with respect to policies after Independence. The desirability of some land reform was agreed upon, and this would be politically feasible because of opposition to the Congress by the largest zamindars, rentiers, and princes. The desirability of economic and industrial development was agreed upon—warmly by the intellectuals and industrialists within the party and less warmly by the Gandhians who favored handicraft industries, but this difference was compromisable. In spite of the disagreement, the Congress had set up a National Planning Council headed by Nehru before the war. There was a general agreement, too, on minimal policy toward the rights of the untouchables and women, the importance of a secular state, and language as a basis of state reorganization. All these were accepted as general policies with many unanswered questions on detail. Since the Congress was not governing, it was possible to adopt policies without worrying unduly about implementation or consequences.

---

[12] See N. K. Bose, *Selections from Gandhi* (2nd ed.; Ahmedabad: Navajivan Publishing House, 1957), pp. 93–100; J. Bondurant, *Conquest of Violence* (Bombay: Oxford University Press, 1959), pp. 63–65; P. J. Rolnick, "Charity, Trusteeship, and Social Change in India," *World Politics*, XIV, No. 3 (April, 1962), p. 439.

Apart from policy areas, the Congress had developed a style of operating, and it was one that had many of the qualities of the Hindu religion and of Indian society. Certain qualities were called for from the leadership, of which one of the most important was sacrifice, and men who obviously gave up narrow class, caste, or community interests and benefits were honored. Within the Congress there was room for disagreement, which could be settled by compromise and without violence. There was a tradition of compromise within the movement—partly resulting from Gandhi—which reflected the Hindu capacity for absorbing and reconciling differences and the long-standing rural traditions of unanimity and consensus. It also meant that policies and statements might be made and issued unanimously, but these were not necessarily binding for implementation. The Congress was now a reasonably effective, organized political institution on a level extending from the locality to the nation, and it had the experience in government it had gained in 1937–1939. It was relatively well organized, but the factionalism so ubiquitous in rural life also played a major part within it. These factions reflected not only policy differences but personal, caste, class, and community differences, and they in fact provided a method of absorbing new groups. But these differences were bridged by the national leadership and by the goal of independence.

The Congress movement took over the national government in 1947 and became the governing party and administration. I will close this discussion of the Congress by quoting one of India's leading political scientists. He wrote in 1961 but his remarks are relevant for the 1947 period and serve as an introduction to the discussion in Part III of political change since Independence:

> The Congress is more than a party, it constitutes an entire party *system.* The conflicts and alignments within the Congress are of greater political import than its conflict with the opposition groups. The operative political categories in India are factions within the ruling party, organised on different lines and interacting in a continuous process of pressure, adjustment, and accommodation. The true opposition that emerges is not against but within the Congress. In this, the opposition parties themselves play their part at the margin.
>
> . . . It is at the level of factions that the true nature of the Congress and its distinctiveness from other parties is also revealed. For it is at this level that one can see how close the Congress is to society: it reflects all the major social divisions and interest groups. It is also at this level that traditional institutions find entry into the political process.[13]

[13] Quotation from R. Kothari, "The Party System," *The Economic Weekly,* XIII, No. 22 (June 3, 1961), p. 849. This is the fifth article in a six part series entitled "Form and Substance in Indian Politics" that appeared in the April 29–June 10, 1961 issues.

PART III

Political Change in India Since Independence

Long years ago we made a tryst with destiny, and now the time comes when we shall redeem our pledge, not wholly or in full measure, but very substantially. At the stroke of the midnight hour, when the world sleeps, India will awake to life and freedom. A moment comes, which comes but rarely in history, when we step out from the old to the new, when an age ends, and when the soul of a nation, long suppressed, finds utterance.

J. Nehru, Speech on Independence
August 14, 1947

The Congress Party . . . is a collection of saints and sinners. It is bound to be with the party in power. It attracts men of all kinds. But what we must do, those of us who have the responsibility now, is to keep the saints on top.

Lal Bahadur Shastri (Interview)
June 11, 1964

. . . Nehru's government is profoundly non-revolutionary.

Guy Wint, *Observer*
May 26, 1963

4

Change Within the Congress Party

INTRODUCTION

Parts I and II and Appendix A of this study presented both the framework of an equilibrium analysis of a social system and a picture of the social structure at the time of Independence in the light of this analysis. Political independence drastically upset the existing social system. The drastic shift in the sphere of political activity would be expected to have profound consequences on power and policy in the economic sphere as well. One of the purposes of this study is to trace through these economic consequences. But first it is necessary to examine the consequences of political independence for the structure of political power in India.[1]

The first and most direct change was that the British rulers left physically, and with them ended the enlightened despotism they had provided. This departure was total in the political sphere and was almost total in the bureaucratic and military areas of government. The successor government was an Indian government, and it had to be both responsive and responsible to the Indian population and the various pressures and demands of the groups within it.

Accompanying independence was partition—with the departure of the large part of the Moslem population that chose Pakistan, and the inflow of large numbers of Hindu refugees to India. In effect this eliminated the Moslems as a major internal Indian political force. Of great importance for India's future, the decision to accept Partition by the leaders of the Congress was a conscious decision in favor of a relatively strong

[1] The literature on the transition period is of course very extensive. I have found especially useful: V. P. Menon's two books: *The Transfer of Power in India* (Calcutta: Orient Longmans, 1957); and *The Story of the Integration of the Indian States* (Madras: Orient Longmans, 1961), Maulana Abul Kalam Azad, *India Wins Freedom* (Calcutta: Orient Longmans, 1959). Penderel Moon, *Divide and Quit* (London: Chatto & Windus, 1961). *The Indian Constitution.* Hugh Tinker, *India and Pakistan* (New York: Frederick A. Praeger, 1962), chap. 3; and "Tradition and Experiment in Forms of Government," *Politics and Society in India,* ed. C. H. Philips (London: George Allen & Unwin, Ltd., 1963), pp. 155–187. Myron Weiner, *Party Politics in India,* (Princeton: Princeton University Press, 1957). Gene D. Overstreet and Marshall Windmiller, *Communism in India,* (Bombay: The Perennial Press, 1960). A recent book is that of W. H. Morris-Jones, *The Government and Politics of India* (London: Hutchinson and Co., 1964).

central government with substantial powers in the economic and social fields over a highly decentralized and weak central government.

One of the immediate and necessary effects of independence, if a united India were to continue to exist, was the integration of 600 princely states into India. This was successfully achieved shortly after Independence. The rulers of these states were replaced as sources of authorized power by civil officials associated with the central and state governments. Some of these princes were also among the largest landowners, and they overlapped with the large, urbanized absentee owners who had opposed the Congress movement. Both these groups lost power as a result of Independence.

Of the Indian political groups that might have assumed power on the departure of the British, the leaders of the Congress movement were the logical and perhaps the only real choice. But the effect of the war upon the movement was to reduce its breadth, and many of the organized subgroups within the movement had broken away. The Communists had cooperated with the British after 1941. The socialist wing had rejected nonviolence as a method during the war and refused to participate in negotiations surrounding Independence or Partition. Also, there had been a wing sympathetic to Subhas Bose and his program of support to the Japanese, but Bose had broken with the Congress by 1939, and with his death and the Japanese defeat this group had little or no focus or influence.

The Congress that took power was narrower than the Congress before the war. Many of the urban-centered, intellectual, and fervidly ideological groups had left it. Gandhi himself refused to serve in a Congress government; he in fact advocated the disbanding of the Congress with Independence, on the grounds, first, that its mission had been performed and, second, that its leaders should work on the local and village level rather than on the national level. Gandhi's moral influence on India and upon the Congress leadership continued to be very great immediately after Independence, and he provided a strong unifying element for that leadership. However, his direct influence diminished, and with his assassination was largely extinguished. On his death many of his followers primarily interested in social reform in effect abandoned party politics.

The Congress found on achieving power that it had lost substantial parts of its membership; it had lost a large number of those representing various ideological elements. However, their number was made up by an influx of another type of member. The Congress controlled the government which was the source of favors and permits and was responsible for new legislation. Those individuals and groups that had something to gain or lose by political and administrative decisions and policies were interested in closer connections with the Congress, and many entered it. In other words, from a membership centering about "have not" groups the Congress began to attract the "haves"—those who would be affected by the decisions and policies of the government. The process of government also required the ability to compromise and arrange, which is opposite to the aptitudes for agitation and opposition most valuable in a revolutionary movement. Many of those who had led the agitational efforts found themselves with a greatly reduced role in the governing organization. Apart from these factors, personal and factional issues that had been transcended by the agreement on Independence as a goal now became more important, and inevitably policy issues became sources of disagreement. The result was that some of the prewar and wartime Congress leaders went into opposition; at the same time the Congress forced out those who disagreed too sharply by adopting a rule that a Congress member could not also be a member of a group with a separate program. What had been a movement representing the nation became a party representing its members and those who voted for it.

The new Constitution of 1950 with its provisions for periodic elections and universal suffrage all contributing to the same end, signified the end of the transition and the creation of a permanent system of government. With the contesting of elections the support of men who influenced votes became necessary, so a political organization on a more or less permanent basis was required. Criteria were necessary in choosing candidates to support or reject, and those who opposed the candidates chosen risked their membership in the party The very fact of free elections encouraged dissident groups to form parties in the hope that they could win in the legislatures. Thus gradually the movements became parties, and by 1950 a Congress party had been organized with the name, prestige, and many of the same leaders as the Congress movement.

The introduction of universal suffrage raised the number of eligible voters from 35 million in 1937 to 170 million in 1952. The effect of the property and education qualifications of the 1937 suffrage had exaggerated urban strength relative to its proportion in the population. Universal suffrage shifted power back to the rural elements in Indian society, because the votes reflected the population distribution. With the relatively ineffective means of communication to the rural areas, the parties tended to rely on establishing contact with existing groupings—largely castes and leaders of castes—to get votes. Thus elections increased the political role of existing rural groups, primarily the castes. Simultaneously the existing village leaders, who were members of the local dominant caste and cooperated with the government both to deal with local problems and to get favors for their villages, now had to work with the Congress since that was the party in power. Inevitably there was a shift in power from more urbanized groups and leaders to those groups and leaders who were themselves rural or who represented rural groups. With universal suffrage, too, elections became far more expensive than heretofore, and the parties sought to attract support from these groups—largely urban—that could finance an election. Thus new coalitions were formed between rural groups that controlled votes and urban groups that controlled resources for elections.

In the early period of Independence the forces contributing to a strong central government appeared dominant. The problems facing India were so obviously national problems associated with establishing and consolidating the new nation that they overwhelmed local issues. The Congress leadership was a genuinely national leadership, known throughout India for its role in the Independence movement. The most important state leaders were attracted to the Center, and the Center had enough prestige to bring local party leaders and legislators into line. Nehru and Patel controlled the government and the party respectively, and they worked in sufficient harmony together to carry out agreed-upon policies. With Patel's death in 1950, Nehru became the undisputed leader of both government and party. He was largely responsible for the victory of the Congress party in the 1952 elections, and no other leader was the national figure that he was. It was in this period that the centralizing forces in the country seemed at their strongest, and many Indians worried about the decay of state and local influence.

However, any Indian party would have to reflect and represent the many linguistic regions in the country. One of the demands that the Congress had accepted prior to Independence was for linguistic states. After some agitation and in spite of opposition by Prime Minister Nehru, this was in large part conceded in 1956 and further by 1960. The setting up of such states followed the establishment of the Indian Constitution in 1950, which set up a federal system of government, with powers reserved to the Center and to the states, modeled on the British legislation of 1935. The justification of a federal constitution was the size and diversity of the subcontinent. Although the central government was given a wide range of general defense, economic, and industrial powers, the state governments were given control over agricultural policy and taxation as they impinged on the individual peasant. The reason for this was the very wide diversity of agricultural practices and tenures within the country. But the grant of power to the states in that economic sector in which 75 per cent of the Indian population was engaged meant, in effect, that the state legislatures and administrations were responsible for passing legislation and carrying out policies that directly affected most of the Indian population. Consequently the state parties and governments became the foci of local political power, and in the process became independent bases of national political power.

In the rest of Part III, I shall look specifically and in some detail at the working out of the political changes since Independence. This will first mean a close look at the changes within the Congress party, which has governed India with an overwhelming legislative majority since 1947. However, the party is separate from the government, and I shall then examine the relationship between changes in the party and the functioning of the government on the local, state, and central levels. The main question that will be asked is which Indian castes and classes have gained or lost in political power? On the basis of this analysis it will be possible in the following section on economic change to look at the changes in economic policies over the decade, and to relate the gainers and losers from those policies to the changes in political power described in this section.

The discussion in this Part on political change will necessarily be qualitative. The basis for the generalizations are a limited and very scattered number of local village and urban studies, or impressions, and several national analyses of general elections that have been made. From these scattered pieces it is possible to discern trends of political behavior in India; however, the conclusions admittedly lack even the semblance of quantitative precision.

URBAN-RURAL SHIFTS IN POWER

One of the main centers of the Congress movement's power was in educated urban groups, frequently western educated; probably they were generally of high caste, and many were Socialists with some tendency toward Fabianism. Nehru was the leader of this element in the Congress movement, but it included many of the Congress Socialist group, Communists, and followers of Subhas Bose. Most of the latter left the party, and these were replaced by groups interested in the gains from power and administrative decisions.

The disillusionment of the intellectuals with the Congress party is seen at its clearest in Calcutta, though Calcutta is in this, as in other respects, admittedly an extreme. The process there has been described by N. K. Bose, one of India's foremost social scientists:

> [When] independence was gained, and the responsibility of administration and nation-building was entrusted to the Congress Party, the vocal portion of Bengal became quickly critical, because, by virtue of the circumstances of peaceful transfer of power, the Congress inherited an administrative structure, which it tried to use for a new purpose. Its idea became, not to disrupt the status quo, but to build up its 'socialistic pattern' of economy on the foundation of the existing order without any violent disturbance.
>
> In this prosaic task of reformation, the Congress Party, in the opinion of

its critics, has tried to convert every problem of national reconstruction into an administrative problem. . . . [There] is an effort to continually add to new responsibilities, instead of a desire to stimulate the growth of non-official endeavours to any appreciable extent.

The identification of the Congress with the status quo, even if the ultimate intention may be one of using it as a spring-board for reform in the direction of socialism, has made the organization unpopular so far as the city of Calcutta is concerned. The loss of ethical quality in the contemporary endeavors of the Congress in the reorganization of its party machinery, or in the matter of running an old administrative machinery without sufficient proof of desire or capability of reforming the latter, has created a kind of frustration, and even of cynicism amongst those who had made the attainment of political freedom synonymous with the advent of social revolution or moral regeneration.[2]

On the positive side, political power within the national and state legislatures and within the party has shifted to a coalition of members of landowning dominant castes and of the small town middle class combined with the upper middle classes of larger cities, especially the new industrial and commercial groups. This shift in power from urban and intellectual groups to a new rural-urban mix is shown on one level in Table 8, which depicts the change in the occupational background of the members of the national legislative assemblies in the four central assemblies from 1947 to 1962. (The Lok Sabha is the popularly elected lower house of the national legislature under the 1950 Constitution.)

It is assumed that the members in all the categories in the table except "agriculture," and half of those in "social work" are probably urban—whether small town or big city cannot even be assumed. Then, it is seen that rural representation rose from about 15 per cent of the membership in 1947 to about 25–30 per cent in 1952, and to 40 per cent in 1962. If the categories of "law, press, and the professions" are considered to include urban "intellectuals" then there was a decline in the representation of this group from 55 per cent in 1947 to 44 per cent in 1952, and to only about 35 per cent in 1962.

This observed trend on the national level is supported and, in fact, is made possible by the trend on the state and local levels which has been pointed out in various studies. Myron Weiner has examined a large sample of political leaders in West Bengal in terms of their place of birth, and found a trend toward the increasing ruralization of this leadership in the fact that the younger leaders are much more likely to come from rural areas than the older. A. C. Mayer in a district in Madhya Pradesh found a shift in power within the Congress from an urban leadership to an uneasy coalition of rural and urban leaders. In Orissa, Bailey has dis-

[2] N. K. Bose, "Social and Cultural Life of Calcutta," *Geographical Review of India* (December, 1958), pp. 27-28.

TABLE 8
OCCUPATIONAL BACKGROUND OF MEMBERS
OF THE CENTRAL LEGISLATIVE BODIES, 1947–1962
(per cent of total members)

Occupational Category [a]	Provisional Parliament [b] (1947)	Lok Sabhas (1952)	Lok Sabhas (1957)	Lok Sabhas (1962)
Agriculture	6	19	22	26 [c]
Business	8	10	10	11
Law	32	25	23	21
Press	11	8	4	7
Education	8	7	3	3
Other professions	5	5	NA*	4
Services	5	2	NA	NA
Social work	14	17	33	27
Miscellaneous or unknown	11	8	5 [d]	2 [d]
Total [e]	100	100	100	100

Notes:
* NA means not available.
[a] Agriculture includes people receiving most of their income from rural holdings, whether as owners or rent payers; services includes civil servants and military officers; other professions consists largely of doctors and engineers; social work includes most of the women members, full time party workers (especially among the Communists) and followers of Gandhi—for most of these their pay as legislators is their main source of income.
[b] Suffrage was restricted and election indirect for the Provisional Parliament.
[c] Includes ex-rulers and members of their families (about 3 per cent of the total).
[d] Includes those in the NA category.
[e] Discrepancies due to rounding.
SOURCES:
W. H. Morris Jones, *Parliament in India,* (Philadelphia: University of Pennsylvania Press, 1951), pp. 114–121; S. Suri, "1962 Elections," Appendix I (New Delhi: 1962); *Eastern Economist,* XL, No. 2 (January 18, 1963), p. 94.

cerned a gradual movement of the leadership within the Congress from the middle-class, coastal, and more urbanized group, which played the leading role in the Independence movement, to a wider coalition of groups that included rural and princely elements from the hill regions. In Madras both Weiner and Beteille have pointed out similar tendencies toward widening the social basis of party leadership since Independence.[3]

Although Nehru remained Prime Minister from 1947 to 1964, several changes in the chief ministerships of the states during that period sym-

[3] M. Weiner, "Political Leadership in West Bengal," *The Economic Weekly,* XI, Nos. 28–30 (July, 1959), pp. 925 ff.; Adrian C. Mayer, "Rural Leaders and the Indian General Election," *Asian Survey,* I, No. 8 (October, 1961), pp. 23–29; F. G. Bailey, *Politics and Social Change,* Part III; M. Weiner, "The Congress Party in Madurai Town" (Cambridge, Massachusetts: Center for International Studies, 1963); and Andre Beteille, "Politics and Social Structure in Tamilnad," *The Economic Weekly,* XV, Nos. 28–30 (July, 1963), pp. 1161–1167.

bolized the shift in power with dramatic force. For example: in Madras the Chief Minister changed from C. Rajagopalachari, a Brahmin lawyer, to K. Kamaraj, a peasant with little formal education. The former knew English and Sanskrit as well as the regional language, and he was the first Indian Governor-General and a national Congress leader. The latter was an astute local political leader who spoke only Tamil well. Kamaraj was definitely not an intellectual, he was hailed as a "man of the people." This might be compared with John Quincy Adams' defeat by Andrew Jackson in the United States.

In West Bengal, for another example, there have been a series of shifts in power in the Congress since 1947: first, leadership by a Calcutta group of political intellectuals and "gentlemen;" second, an uneasy coalition of Calcutta elements headed by Dr. B. C. Roy, the long-time Chief Minister, a western-educated doctor and a sophisticate from Calcutta; and third, a smaller town-rural group headed by P. C. Sen and Atulya Gosh, who are both from Hooghly district, educated within the state, and connected with Gandhian elements within the Congress. On Dr. Roy's death in 1962, P C. Sen became Chief Minister, and Atulya Ghosh became the unchallenged head of the state party.

In a very perceptive article on the leadership of the Communist party in West Bengal, Ashok Mitra points out that a shift similar to that in the Congress party has also occurred in the Communist party in that state.[4]

CASTE IN POLITICS

How have these shifts in power between rural and urban elements related to the trends in Indian caste and communal structure? What is the role of caste and ethnic groups in current Indian politics? The answers to these questions are complex and varied, depending in part on specific areas. First, caste is the major principle of social organizations in rural India, and community, expressed as a linguistic factor, is extremely important in urban areas. Since the rural vote is playing a much greater role in politics than heretofore, and since the demand for independence, which largely transcended caste or area politics except among the Moslems, has been resolved, it would be naive to expect that caste and ethnic elements do not play a greater role in Indian politics. They play a part no less than religion and place of national origin play a part in American politics, especially on the state or local level. Second, within the constitution, about 15 per cent of the seats in the state and national legislatures are assigned to members of the scheduled castes, which are largely the untouchable castes listed in the Constitution.

---

[4] A. Mitra, "The Plebian Revolution," *Seminar*, LI, November 1963, pp. 33–36.

The exact role caste factors play in any particular area of India cannot be predicted on a priori grounds. In purely rural areas it is probable that they play a larger role, since there are few counteracting factors. Whether caste is a major factor within local rural elections depends on how strong caste conflict is in the nonpolitical areas of rural life; it will likely be important if a lower caste is seeking to move up politically to reflect or to support possible economic and social gains it has achieved in nonpolitical areas, or if the existing dominant caste is in such firm control that it is not challenged and can control the local elections. Bailey presents one interesting case of a relatively isolated village in Orissa in which the lower castes had advanced economically, and the caste factors in local elections were very strong; but in another village, near the capital of the state, caste was blurred by a great many other interests, and a relatively large middle-class group of intermediaries not confined to a caste had arisen.[5] As a result, in this latter village, caste elements played only a minor role in elections. Perhaps equally important, the members of the scheduled castes were so depressed economically that they had neither initiative nor expectations. Even the Communist party had given up any attempt to appeal for their votes on a caste basis because of their political apathy.

Probably the first point to stress in a detailed analysis of the role of caste factors is, that the role of the subcaste or *jati* elements in the social structure is limited by the geographical extent of these groups. The subcaste—with consistency in ritual rank, economic strength, and political power—is a local element extending over a small area of intermarriage or, at the very widest, over a linguistic area, in which case the recognition of its name and rough status may be state wide. This means in effect that, with universal suffrage when a constituency area for state and national elections is always larger than a single village or cluster of villages, the support of only one subcaste is ineffective: there must be a wider basis of support. This may of course also be true within a single village if no subcaste has a majority of the votes, or if the dominant caste is split into opposing factions. In such cases it is necessary to win the support of several factions, castes, or groups rather than just one, and then the problems of coalition arise. It is also clear that the larger the number of voters that a caste or faction can control within a constituency the more desirable it is for a candidate to have its support, assuming its support does not give rise to a coalition of a majority against it. Various alliances of subcastes or castes have in fact developed. Mayer uses the term "allied castes" to describe castes of roughly similar status and eco-

[5] F. G. Bailey, *Politics and Social Change,* Part I.

nomic position, although of different occupations, in an area. These allied castes of high rank supported each other politically in the region Mayer investigated.

Of even greater importance than traditional caste factors, on the state or national electoral level, is the emergence of caste associations into a political role. As pointed out earlier, these caste associations were started in many cases as voluntary organizations of people of the same general caste for setting and enforcing caste rules, frequently as part of a sanskritization process. In the process they set normative rules of behavior that were of greater ritual purity than the existing practice, and they frequently operated schools and welfare societies for their members, or lobbied for their members in relations with the government. They are generally statewide and have a permanent bureaucracy that consists normally of an educated and aspiring leader supported by a small staff. These associations are an obvious target for political parties wooing mass support, and they themselves see politics as a method of improving the position of their members in the caste hierarchy, especially since one of their major aims is to improve that position in relation to the government. The Rudolphs have described the role of caste associations in India, but with emphasis upon Madras, and they pointed out that the caste association provides a traditional framework that makes the new democratic processes in India comprehensible to the politically unsophisticated masses.[6] But these caste associations vary in strength from state to state. In Orissa and Rajasthan, among the less developed states, they are weak; in such states as Madras and Kerala, they are strong. Madras is developing rapidly, and both have a high rate of literacy; jockeying for position among the castes is a characteristic of their social life. The caste associations are very influential in economic, social, and political areas in those two states.

In some areas scheduled caste associations have been organized for both social and political purposes, especially in light of their special political and economic rights. Some of these were originally started by higher caste individuals, who were influenced by Gandhi and acted on ethical grounds to improve the position of the lower castes; others were started by the Harijans themselves. Probably the most important political organization started by the scheduled castes on a wider level is the Republican party. Although this is a national party it is largely confined to the Harijans of Maharashtra and Central India, many of whom left Hinduism

[6] L. I. and S. H. Rudolph, "The Political Role of India's Caste Associations," *Pacific Affairs*, XXXIII, No. 1 (March, 1960), pp. 5 ff.; see also M. Weiner, *The Politics of Scarcity*, chap. iii.

and became Buddhists. The party was organized by Dr. Ambedkar, the Columbia University graduate who became the leader of the scheduled castes and one of the main authors of the Indian Constitution. In contrast to this independent party, the scheduled Chamar caste in North India has been won to the support of the Congress party. In Kerala one of the main supports of the Communist party is the low-caste Ezhavas, who have been rapidly advancing economically. And one of the main reasons for the downfall of the Communist government in Kerala was the departure of the strong, upper-caste Nair Society from its supporting coalition on the grounds that the government was favoring the lower-caste Ezhavas in its educational policy.[7]

In some states the statewide caste groups and their associations are intimately connected with the political parties, so that the parties and their factions reflect the caste conflict. In Andhra the conflict between Reddis and Kammas—the two major landholding castes—has been closely involved with state politics; the Reddis have been supporting the Congress party and the Kammas, at times, the Communist. The existing opposition and the Communist policy in Andhra have reflected an effort to retain landowner support.[8] In Mysore the conflict between the Okkaliga caste, which consists of about 15 per cent of the present population though it was a larger part of the smaller population before 1956, and the Lingayat caste, which now forms about 20 per cent of the population, is reflected in the battle for power within the state Congress party. In Mysore about 18 per cent of the population are Harijans, who until now have played a negligible political role in the state.[9]

The rise of the Swatantra party in both Madras and Mysore in the south is closely connected with caste issues that center on the decline of the Brahmins.[10] The Brahmins are the mainstay of the Swatantra party in Madras. Although non-Brahmins have won control in the Congress, it is still a coalition with some Brahmins. Finally there are the Anti-Brahmin, Anti-North parties—especially the Dravida Munnetra Kazhagam (DMK) —which are significantly the main opposition in Madras to the Congress

[7] See on this "Role of Caste in Determining Party Loyalties," *The Statesman* (Calcutta), May 18, 1961.

[8] But the Congress split the Kamma support for the Communists by itself nominating Kamma candidates in the districts in which the Kammas were numerically strong; and by appealing to the property interests of the landed Kammas. Thus caste differences were subsumed by class ones to an extent sufficient to defeat the Communists sharply. On Andhra see S. Harrison, *India: The Most Dangerous Decades* (Madras: Oxford University Press, 1960), especially chap. vi.

[9] M. Srinivas, *Caste in Modern India* (Bombay: Asia Publishing House, 1962), chap. i. Mysore was enlarged in 1956.

[10] On the South and the DMK, see especially A. Beteille, "Political and Social Structure in Tamilnad," *The Statesman* (Calcutta daily), Sept. 18–19, 1962.

party. Still, both the Swatantra and Congress parties have cooperated with them at times. The supporters of the DMK and similar Anti-Brahmin parties consist of relatively urban, lower middle class groups such as students and clerks in the salary range of Rs 200–350 per month or some labor groups, especially taxi drivers. They get financial and publicity support from the new Tamil-language film stars. In some districts of Gujerat in the northwest, the rise of the Swatantra party is closely associated with the economic improvement of landowning castes in the area. These castes have in the past been strong in number although weak in economic strength, but they are now challenging the position of the smaller and economically dominant trading castes traditionally associated with the Congress party.[11]

What does become clear from the previous series of regional vignettes is that castes and caste groups are extremely important in politics, but, except possibly in the smallest village, larger bases of support must be built up from collections of castes of a similar character united either for or against a program or person. And, in fact, the success of a party lies in its ability to transcend individual caste groups. This is one of the major reasons for the political success of the Congress party. Specifically it tends to have the support of scheduled castes, though this may not hold in every state. This support reflects the influence of Gandhi, but even more it reflects the steps the Congress has taken to improve the condition of the scheduled castes—land reforms, reservation of educational and political opportunities, cabinet seats, and the like. Because it is both a national and a secular party and because its long-time leader, the late Prime Minister Nehru, was the nation's foremost advocate of a secular state, the Congress has also benefited from the support of minority religious groups or communities, Christian and Moslem, who fear the less clearly secular parties. In Kerala, where the large Catholic population plays an important role in the Congress, the Catholic influence also has some disadvantage in encouraging Hindu support for anti-Congress parties. Part of this support also comes from the traditional attitude of minority groups in India to look to the government for protection and favors, and the Congress is identified with the government today. This minority support on a nationwide scale together with the support of sufficient other caste or noncaste elements provides the basis for the Congress party's national appeal and makes it in fact the only strong national party in India. The individual opposition parties may be almost as strong as the Congress in

---

[11] On Gujerat: Rajni Kothari and Ghanshyam Shah, "Caste Orientation of Political Factors," *The Economic Weekly*, XV, Nos. 28–30 (July, 1963), pp. 1169–1178.

specific districts and even states, and in a local election they may be able to defeat the Congress by combining in opposition; but they have not been able to combine their strengths or to agree on policy in a national election. Therefore they lose, although the Congress does not have a popular majority.

Thus we may summarize this discussion of caste: first, it is a traditional method of organization; it has always been associated with political power, and it is neither surprising nor to be unduly deplored that it still provides a means of organizing power in democratic Indian politics and in rural India especially. Second, the smallness of the subcastes and their very local character make it necessary to achieve a wider base to win elections in a constituency. This means that coalitions must be formed in many areas, and the power of any single subcaste consequently becomes diluted. Where factions exist within castes in a village or small surrounding areas the same holds true. The need to form coalitions has permitted a broadening of power within rural India.

The caste groups that were small in number but dominant in an area before Independence because of their economic or ritual status or because of their close connection with the British, have lost power. This reflects both the political reform of universal suffrage, which made numbers one of the key elements in political power, and various economic policies that distributed economic power more widely than previously. At the same time the active role of the government in economic development and the fact that economic policies in the form of development expenditures, allocations, and changes in land rights—partly determined at the village level—themselves were felt as low as the village level, and made political power worth fighting for. This has occurred increasingly and has distributed power more widely among the populous small-landholding groups and peasant castes that are both aware of the gains to be derived from political power and have enough economic strength to be independent. This shift has also meant the loss of some power to the former princes, large zamindars, and members of the Brahmin castes. These groups, however, have resources of traditional prestige and funds, and thus have recovered some of their initial losses through their own political action in the Congress or opposition parties. In some areas the traditional, dominant, non-Brahmin castes have been able to maintain their economic and political power, either independently or in alliance with other castes; in other areas they have lost some power.

Since wider support than that of the subcaste has become crucial politically, caste associations have been playing a larger political role than heretofore. This political role fits in with their assumed role as agents of sanskritization and their methods now extend to include political means, because control of government jobs and access to education, which are signs of higher status, are political decisions.

Caste becomes a major issue in local or regional politics when it reflects other changes. If members of a caste are either striving to improve their political position to reflect the change in their economic and social positions or are aware that an improved political position will influence the latter, then caste becomes an issue. Likewise the reverse holds true: if members of one caste are threatened by another rising caste, then caste will be an issue in an election. But if dominance is unchallenged, as it is in many areas, or if lower castes are apathetic, then it is not an issue. Similarly, the more the problems of a village or rural region are tied in with the problems of a larger area and the more a middle class develops that links the locality with the larger area, the less of a role do undiluted caste factors play; because in the larger area caste is modified by many other influences.

In my earlier discussion of caste I pointed out that scheduled castes, largely the untouchables, have no way of rising within the orthodox Hindu system. This means that they must rise through political means. In the course of their political efforts in Uttar Pradesh and Punjab in the north, they have worked through the Congress party, and this has been at the expense of strained relations with their landlords, high caste Jats, who supported the Jan Sangh, partly because of the Congress attitude toward untouchability. In other states they work through opposition parties, most notably the Republican party in Central India and the Communist party in Kerala. But in all these areas the untouchables are improving their economic position and are striving for the social rights granted to them under the Constitution, and one part of this effort necessarily takes a political form. In areas in which they are not advancing economically they are politically apathetic, and their role in politics as an independent force is negligible—though in a subordinate role they provide a "vote bank" for their economic and social superiors.

Caste elements probably play a greater internal role within the Congress party than in other parties although these latter parties may represent caste groups in opposition to the Congress, and the Congress ideologically is a secular party. The reasons for this paradox are simple. The Congress has been in unquestioned power since Independence and it would be natural for political conflicts to take place first as factional fights within the Congress before losers depart. Losers may and frequently do reenter the Congress party at a later stage. In many cases the opposition consists largely of the members of one caste, and therefore within it there would not be an intercaste conflict, although it may contain factions. The members of the dominant caste in an area will probably be the leaders of the Congress in that area. Weiner and others have noted that the larger landowners in districts in Madras State, in Mysore State, and in Maharashtra are leaders of the Congress party; Weiner has also pointed out that Congress leaders in West Bengal are more likely to be leaders of caste and social organizations than of class groups. The secular opposition leaders, especially the Communist and Socialist parties, are more likely to be leaders of class organizations set up by those parties, which stress class as a force. Since those parties have attracted the urban intellectuals, they are more likely to be from areas in which caste plays a less important role. When they get into power, as in Kerala, caste may play a major internal role in their policies also.

It is one of the major strengths of the Congress party that it does represent and absorb various traditional elements of Indian society. Weiner has pointed out that the Congress party specifically is strong in performing those political functions—expediting by providing links between the government and citizen, constructive and public service, and mediating among groups—that the members of the dominant castes in the villages and regions have always performed; and those members, or members of new dominant castes that have greater numbers than the old, are now the leaders in the Congress party and in the new government institutions in their villages, districts, and states.[12]

NONCASTE GROUP ELEMENTS IN POLITICS

COMMUNITY

One of the most important factors transcending caste and making it possible to transcend caste in politics is community. Community is essentially determined on either a linguistic or a religious basis; and on issues that are considered to have a vital effect on the community, all subgroups within it, whether of caste or class, unite. It was such unity on the issue of the creation of linguistic states that forced the Center to yield to the demands of the Telugu-speaking population for the creation of Andhra. In Bombay State the demands by the Marathi-speaking people of the southern areas and the Gujerati-speaking people of the north led to the actual splitting of Bombay into two states with Bombay City going to Maharashtra State; but this was done only after the Congress party was seriously threatened politically in both states by coalitions of parties fav-

---

[12] This point is made in M. Weiner's mimeographed paper, "Traditional Role Performance and the Development of Modern Political Parties—Reflection on the Indian Case" (Massachusetts Institute of Technology, Center for International Studies: 1964).

oring the split. In West Bengal the imaginative idea of unification of Bihar and West Bengal was politically defeated by the unified opposition of the Bengali-speaking population in a district by-election. In all these cases the communal differences between the intrastate linguistic groups were supported by class differences between the communities. For example, in Bombay City the more numerous Maharashtrans were largely working class, although the incentive and leadership in the movement were provided by intellectuals and other white-collar elements; the Gujerati were largely middle class and in control of many of the local businesses. In such cases the communal groups were large enough within their areas to take over the local Congress parties on this issue and force the national party to accept the desired aim. In the case of the demand of the Sikh community for a separate state within the Punjab, however, the Sikh community is not large enough or so compact in area, in relation to the Hindu community, to take over and change the Congress party policy; similar situations exist in the case of small marginal boundary disputes between states, where the local communal issues become subsumed under much larger problems and do not result in a communal solution.

But in urban areas, especially, where there is a mixture of communities the element of community plays a role in urban politics somewhat similar to caste in rural politics.[13] Candidates are selected on the basis of their community in relation to the constituency, and it is important for a party to develop a balanced community list. However, although each party may seek to win a community by nominating a member from it for office, the fact that the community is normally large makes it possible for competing parties to nominate candidates from the same community; then the election is settled on the basis of other issues. There are important communal differences in voting, like the tendency (I have mentioned it before) for minority groups to favor the Congress. In Calcutta the non-Bengali-speaking districts of the city—largely populated by people from Bihar, Orissa, and Uttar Pradesh—tend to show a relatively greater vote for Congress than the Bengali-speaking districts. But these differences are as much, if not more, on class issues than language issues. The Bengali-speaking, middle-class population has been anti-Congress, left or right, for a variety of reasons not related to language;[14] whereas the working class members, with roots still in their rural districts, support the Congress party that they know from their rural background.

[13] H. Hart, "Urban Politics in Bombay," *The Economic Weekly,* XII, Nos. 23–25 (Special Number; June, 1960), pp. 983–988.

[14] N. K. Bose, "Social and Cultural Life of Calcutta," pp. 27–28.

CLASS

A factor related to community and caste but cutting across both is the class element in politics. I have defined class in economic terms measured by level of income or control of property, although it is also related to aspirations. To a certain degree caste and community are related to occupations and class, and I have already pointed out these relationships; but class also cuts across caste and community, and may in fact be more important in determining position.

In the rural areas land reforms and government programs have had a definite effect upon the class structure. The princes and largest landowners have lost some power; they may still have local prestige and loyalty, which they can use to contest elections, but they no longer absolutely control votes. Those who remain as landless laborers are dependent upon higher castes and are often politically apathetic. But the middle-landowning group, usually the landowner who cultivates his own land, is now a political force to be reckoned with, for it is playing an increasing role in village elections under the "panchayat raj" program, in state politics, and in the national party organizations. This last group is challenging the traditionally dominant caste groups in some areas, and although it tends to support the Congress since the Congress party is responsible for its gains, it may also support other parties—as the Swatantra party in Gujerat—if the Congress party is too closely identified with the challenged dominant caste. Similarly, the ambitious peasants with smaller holdings and the agricultural laborers in the North who are in the untouchable castes, have opposed their landlords. This conflict has been taking the political form of a struggle between the Congress party, with lower class support, and the Jan Sangh, with landlord backing.

In urban areas working-class organizations such as unions are affiliated with each of the major parties. As I pointed out earlier the leadership of these is most likely to be middle class. In the Congress party the unions do not generally play a major political role: the case of Ahmedabad would be an exception, but the Ahmedabad Textile Labor Association, founded by Gandhi, is an exceptional organization. In part, this minor role reflects intra-union squabbles, but more important it reflects the greater influence of other groups in the party and the peculiar role of the Congress party unions as instruments of the government. There is also at least some desire on the part of the Congress leadership at the state level to prevent the introduction of union leaders into party positions of influence.

The trade unions are far more important in the Socialist and Communist parties because both parties have a narrower membership than the Congress, and because they supposedly represent the working class. Both parties have unions affiliated with them, and they are far more likely to nominate trade-union candidates for office in urban areas than the Congress party. And in working-class districts a candidate's devotion to the working class is a source of political strength. Still, unions are not politically strong in India. The reasons for this are complex, and I shall simply indicate some: the relatively small proportion of the population engaged in factory industry; the rural background of the workers, and the fact that they maintain rural interests; the problem of leadership, and the middle-class character of the leadership; the peculiar role of trade unions in a developing country and their close contact with the government; the government's great role in labor relations. All make the unions relatively passive in that they are acted upon by political forces, rather than being politically initiating groups.

The middle class plays a much greater role than the working class in politics. This was the case prior to Independence in the Congress party; and though middle-class groups are still important, their role has changed. First, the urban intellectuals, one of the middle-class groups that was most important in the Congress party before Independence, have in large part left the Congress party. They still play a role as critics, as journalists, and so on, but their role in the party is relatively minor. They now play more important roles in the Socialist and Communist parties.

Another group—the lower middle-class clerks, teachers, and the like—also have a different role now. Those who started out the 1950's in this class have probably improved their economic condition to only a minor degree, if at all, and have lost relative to other groups. The members of this class, especially when their class membership is associated with caste position, have supported opposition parties, frequently shifting from the communal extreme to the Communist extremes. However, there are many new members of this lower middle class. In the rural areas there are the new private intermediaries between the village and the city or the government, and there are also the numerous new government officials who are administering the agricultural, education, and health programs. They are often not members of the Brahmin castes; in fact, in many areas they are members of peasant castes, even of scheduled castes. In urban areas there are the new clerks in factories and private and government offices, and again they are often members of lower castes. These groups have gained by economic development; and although their gains may not attain their expectations, they are likely to support the party that was responsible for the new jobs.

Finally, there is the upper middle class. Within this are the growing and the new industrialists and managers primarily in urban areas but also in rural areas. This is one of the groups that has gained significantly in economic terms from the industrial development that has occurred, especially in the private sector. Its members are deeply involved in government through the network of administrative mechanisms and institutions that are responsible for grants of licenses, foreign exchange allowances, tariff protection, setting of prices, and allocation of scarce raw materials. At the same time the political parties are interested in them since the costs of election campaigns are substantial. Very little is known about actual campaign costs, but at present the legal limits of election expenditures by a candidate are Rs 25,000 for a Lok Sabha seat and Rs 7,000 for a seat in the state assemblies.[15] Within these limits the cost to a single party of campaigning for approximately 500 Lok Sabha seats and 3,000 state legislative seats would be on the order of Rs 32.5 million. In fact, this is an underestimate. A. D. Gorwala, a very knowledgeable ex-administrator, estimates a total cost of about Rs 50 million for the Congress party to conduct its campaigns. Even this is probably low; the actual cost to a party is often more in the neighborhood of two to three times the legal limits, which would make the total cost to one party of contesting all districts about Rs 75 million (or $15 million).[16]

Much of the Congress party's finance of these elections comes from the large industrialists. It has been estimated that in the 1962 elections various major units of the Tata group contributed approximately Rs 1.0 million to the Congress party; the contribution of the Birla companies was approximately the same; a major cement company contributed about Rs 500,000; and other companies gave substantially. The Tatas and the cement company were also major contributors to the Swatantra party.[17] Many of the industrialists are also large contributors in a personal capac-

---

[15] "Comment," *Opinion*, IV, No. 5 (June 11, 1963), p. 2.

[16] *Ibid.* In the last elections it was estimated that the cost per assembly candidate in Andhra State for the 1962 elections was approximately Rs 50,000. In Calcutta the direct cost for an assembly candidate has been estimated conservatively at Rs 25,000, and for a member of the Lok Sabha at about Rs 100,000. In the Orissa elections for the State Assembly in 1962, the alleged cost per seat for the Congress party was on the order of Rs 25,000. This estimate may be unduly high, but it is alleged that the cost per seat for the leading opposition party was at least Rs 15,000: we might take an estimate of the same amount per seat as reasonable for the Congress.

The Andhra figure is given in *The Statesman*, February 10, 1962; the Orissa figures in *The Statesman*, June 14, 1961, and the *Times of India*, June 16, 1961; the Calcutta figures were given in private conversation.

[17] Figures given in "The Malaviya Affair," *The Economist*, CCVII, No. 6243 (April 20, 1963), p. 227. These figures are unquestionably based on the published annual reports of the companies concerned, since political contributions by companies must be published in their reports.

ity. These contributions are perfectly legal, and are expected by the Congress as well as other parties. In exchange they grant the contributing industrialists access to the positions of power on either the state or national levels and guarantee a hearing at least on a high administrative, and possibly political level, on their more or less inevitable and continuous appeals to the government for necessary permits. Thus these contributions may be regarded as a necessary cost of both entry into industry and conduct of business. At the same time the close connection between industry and the Congress creates somewhat of a vested interest by business in the success of the Congress party, with which connections have been established since it has been the governing party; and failure to support the governing party financially can cause reprisals that may seriously influence the future of a businessman. Financial contributions, however, do not insure decisions favorable to one contributing group. There is strong competition both among the largest groups, and between local and national business groups. I have also pointed out the social and ideological bias against big business among both party leaders and top administrators, and there is an inquisitive Lok Sabha membership that may raise embarrassing questions. All these factors diminish the role of financial contributors between elections, without denying the influence of the givers, which is greatest in its effect upon administrative decisions in the implementation of broader policies. With respect to general policy, it is probably greater as a negative or delaying role on a particular issue than as a positive force in the making of new policies.

The new professional groups are another part of the middle class that have supposedly made large financial and publicity contributions to various parties. The movie stars in South India are a major source of finance and public relations support to the South Indian, anti-Brahmin DMK party. Local industrialists may also support this party. In the last elections in Bombay in 1962, a wealthy doctor was the head of Krishna Menon's campaign, and Mr. Menon received support from various movie stars. His opponent probably received support from industrialists.

Very little is known specifically about the contributions to other parties. The Swatantra party receives funds from large industrialists in Bombay, especially from Parsee groups. It is probable that some of the more orthodox Hindu industrialists have contributed to some of the Hindu communal parties, and that well-to-do intellectuals and professionals are among the contributors to the various Socialist parties, the non-Communist left, and possibly the Communist party; but there are no details.

Although class factors play a part in politics, at times in support of caste and communal factors and other times cutting across them, the importance of these class factors varies. In the area of party finance they may be of great importance; in fields of choice of candidates, general policy and issues, and as vote catchers, they are of less independent importance than other factors—such as caste and community—with which they are intertwined.

PERSONALITIES, POLICIES, AND IDEOLOGIES

A major element in the choice among candidates is the personality of the candidate. Where the group factors are weak or cancel out, this may be the major factor and override group elements. The Indians are very conscious of an individual's membership in a group and are skeptical of his actions apart from the group, but a person who has proved his willingness to transcend group motives by actual sacrifice is very highly regarded. The examples in recent history of Gandhi, Nehru, and Patel are too well known to need further discussion. A candidate who has proved his sincerity in this respect has a good chance of nomination and election, even if he is from a numerically weak caste, because he overrides caste.

Under the British with the district officer making decisions on the spot, and under the princes with the ruler available for decisions, authority was of a personal character. The political strength of the princes, or their heirs even today, lies in the fact that a princely candidate represents a person to whom the villagers feel they owe an obligation and from whom they feel they can get a response without bureaucratic delay. The Maharani of Burdwan can run on the Congress ticket, or the Maharani of Jaipur on the Swatantra ticket; they appeal to their constitutents as a mother to children, and both win overwhelmingly.

Quite apart from this, personality plays a major role in Indian village society. Conflicts in village subcastes, in larger caste or communal groups, and in parties often revolve around conflicting families, personalities, and ambitions, and this gives rise to factional disputes. Again, in areas where broader policy issues are not clear or are not understood, local factors and the persons representing them may be of crucial importance in elections, and are stressed in the elections at the expense of policy. This is inevitable in both developed and underdeveloped countries but probably plays a greater role in the latter because of the role of family ties in all of life, as well as such technical factors as restricted communication and education, which narrow the voter's understanding of broad issues.[18]

Within the national Congress party, with all its factionalism, the late Prime Minister Nehru had a role above faction and even above party, and he won that role because of his great personal appeal to the Indian

[18] Namier's general description of English politics in the eighteenth century are relevant as well for India today.

people. He was even able to use that power just before he died, in spite of age, illness, and increasing criticism, to achieve certain ends that he wanted within the party and government—to get the resignations of a group of leading cabinet members and chief ministers of states. And although he had to compromise on many issues, he was in large part responsible for the secular and socialist character of the Congress party and government today. He in effect represented the westernized, intellectual group within the country. He pushed such policies as economic planning, the socialist pattern of society, and coöperative farming in the economic field; while in the field of social policy he was one of the strongest representatives of the secular forces in Indian life and advocated loosening the traditional restraints of caste and religion.

Within the Praja Socialist party, the departure of Jai Prakash Narayan from politics cost that party its most widely known leader and contributed to its weakening. And Narayan was until recently considered as a possible Prime Minister precisely because of his national appeal, even though he is not a Congress party member.

The appearance of the Swatantra party is closely allied with C. Rajagopalachari's return to politics. Though he was repudiated by the Congress party in his home state, the former leader of the Congress and the first Indian Governor-General has great national appeal—far more appeal than any of the regional leaders who play active roles within the new party.

In West Bengal, Dr. Roy played a role before his death vis-à-vis the state Congress party similar to that of the Prime Minister in India. He represented a force above party, and he advocated policies and governed on a personal level, which appealed to groups that might not support the party itself. This was especially important in Calcutta where Dr. Roy influenced the urban intellectuals who, without his appeal, might have supported opposition parties to an even greater extent than they did.

It is significant that these were all leaders of an older generation—from the period of Independence. And all these leaders and the personal element that they represent in politics are, in fact, becoming less important vis-à-vis the organizations that are required for governing a country. It is certain that the new leaders of India will be different from these older ones who bridge the heroic age of Independence to the earthly age of self-government.

Some of these new leaders have already emerged in the Congress—such men as Lal Bahadur Shastri, Nehru's successor; Kamaraj from Madras; P. C. Sen and Atulya Ghosh in West Bengal; Chavan in Maharashtra; Sanjiva Reddy in Andhra; and Patnaik in Orissa. None has had

Nehru's national strength. All have been closely connected to and derived their strengths from the Congress organizations in the country or from their states; have been considered good administrators with ability to delegate authority; one at least, i.e., Patnaik, moved from business into politics; none are even remotely the westernized, intellectual, high-caste leaders of the Nehru or Rajagopalachari types.

Finally there are national issues and policies that transcend group factors. It is often claimed that these are not important in Indian elections, but that is not true. An obvious example was Independence. The non-Brahmin groups in South India, who had supported a strong pro-British "Justice" party based on anti-Brahminism, lost sharply to the Congress party in the elections of the mid-thirties; but after this loss many joined the Congress and eventually dominated it in the South. Subsequent to Independence, the Congress party gained greatly because its leaders were most identified with winning it. This opposition to imperialism and the desire to maintain political freedom and independence have strongly colored Indian foreign and domestic policy since.

In the early stages of Congress' thinking, the desirability of economic planning for development was not an issue: the few groups aware of the problem agreed, both for social reasons and for reasons of political independence, on the need for planning even though there were some differences on the aims and methods. But only a very small proportion of the population could have been aware of the issues. Today, because of the economic achievements and discussion of the 1950's, the desire for economic improvement on a personal and national level is apparently deeper and general;[19] the Congress by its policies and achievements anticipated that desire. To a certain degree if the Congress had any policy identification and program in the first two elections, it centered on its reputation as the leader in the fight for political independence and on its program in the Five Year Plans, which are as much political as economic documents. The Congress party has also been successful in taking over policies of smaller parties, such as the "socialistic pattern of society," and has won support thereby. One of the main areas of policy disagreement between the newly formed Swatantra party and the Congress party is the opposition of the former to planning on both ideological and administrative grounds. To some extent the role of opposition intellectuals, whether on the left or right, is to develop opposition policies, which in some cases have been taken over by Congress. But it is doubtful how much influence these issues have in fact had in elections; and they are

---

[19] On this see A. H. Cantril, *The Indian Perception of the Sino-Indian Border Clash* (Princeton University Press, Princeton Institute for International Soviet Research, 1963), especially pp. 1–15.

almost never posed in terms of specific choices among policies for development.

These broader policy issues, transcending caste or communal factors, may be of some importance in the urban areas, but they are of very minor importance in rural elections. In rural areas specific policies centering on land reform, land taxes, and agricultural policies are major issues, and here again there are differences between the Swatantra and the Congress parties. One of the important problems in this area, however, is to determine what is the party policy. The Congress party, meeting in general session, may agree unanimously and publicly on a policy of coöperative farming, for example; in fact the unanimity is superficial and may simply reflect an unwillingness to oppose the Prime Minister, for the state government—responsible for any legislation—and the state parties may not implement the policy. In rural areas also, the Congress party is identified with programs to improve the position of the untouchables, and it wins and loses votes on this issue. But in both urban areas and rural areas the Congress party does have an image, which is based on the personal leadership of the late Prime Minister Nehru, on its past relationship to Mahatma Gandhi and Independence, and on its present power. Some final part of the image is based on its broad national policies: economic development based on planning stressing industry; land reform; against the "feudal" elements and in favor of the lower castes; and a secular approach to politics, which was closely related to Nehru. This image may be as unclear and as internally contradictory, especially on the state and local level, as the images of the Democratic or Republican parties in the United States are to a logical Frenchman; but as the Democratic party attracts votes from Negroes and labor as well as from southern conservatives, so the image of the Congress party in policy and personal terms still attracts votes and enables it to win overwhelmingly on a national basis.[20] At the same time the party does reflect the traditional forces of caste and community in Indian society, both in terms of their inherent strength and their broader—even diluted—roles in a democratic environment. The Congress party does this especially in its role as an intermediary to the government, and it is this combination of new and old that contributes to its strength and past success.

[20] In fact, the Congress party resembles American parties in some respects, especially as a loose coalition of groups—far more than it does European parties—and this is one of its strengths. Its opponents are far more logical in policy terms in part because they are not in power.

5
The New Role of Government

In the previous chapter I discussed the changes in political parties since Independence and showed how they largely reflect the introduction of universal suffrage and self-government to India. Meanwhile, the role of government expanded greatly, mainly because of its extension of responsibility to new areas of activity—primarily economic areas. The Congress party, which has won all the national elections and most of the state elections, has appointed the ministers of the government. But the party is not the government. When the ministers enter the cabinet and when the legislators are elected to the legislatures, they have responsibilities to their party though it represents only a portion of the population. In fact, they may not represent their party in any meaningful sense, but only a particular faction within it. Thus the Congress party is in one sense an especially potent source—among a variety of such sources—of pressure upon the government; but it is separate from the government, and conflict may arise between it and the government.

The main interest of any party is to reëlect its candidates, retain control of the government, and survive. The elected members of the government generally hope to be reëlected, or if not, then they hope that a member of their party will replace them if they decide to retire. They must therefore be in close relation to their party. At the same time, the government has a far wider set of responsibilities since it has a far wider range of functions. Furthermore, the elected members of the government who are also active party members are only a small part of the total number of government officials. Under a parliamentary form of government, ministers are constitutionally responsible to the Parliament for the acts of their subordinates. The members of the legislature who are not ministers are affected by the acts of the departments headed by the ministers, and their interests may be quite different, even though they are of the same party and may support the ministers in a vote. Finally, there are in India, as in all countries, very many more or less permanent members of the bureaucracy subordinate to the political ministers who head the departments. They themselves often are not affiliated with a party, and though they supposedly carry out the policies of the ministers, in so doing they inevitably make policy to some degree.

The most important fact of government of this past decade is the great widening of the area of government responsibility in the field of economics. This is in sharp contrast to the former British government's peacetime policy, though not its wartime policies. The goal now is a new one: general economic growth rather than military expansion and efficiency. The effect of this has been that the government, at one level or another, is much more intimately involved than before with the detailed economic activities of all types of individuals—those of low as well as high status. These government activities and decisions mean the difference between success and failure in the economic plans of individuals and groups of individuals. The volume of government expenditures, the decision to put a factory or a dam in one area or another, the decision to allow the building of one type of factory or another, the grant of a license to set up a factory or to get the foreign exchange to construct a project, the program of taxes, the policy toward land reform and consolidation—all affect the economic well-being of individuals and areas, both absolutely and relatively. Consequently, to some degree or other, these government actions will effect their political preference and votes. Economics is today the stuff of government in India and, as such, the stuff of politics on all levels.

VILLAGE AND RURAL DISTRICT GOVERNMENT [1]

There is a major recent change in the institutional framework of village government: the elected village panchayat, formerly more or less powerless and pointless, is now the lowest rung of a hierarchy of directly or indirectly elected bodies—village, development block, and district. They have definite responsibilities for planning projects on the village and district levels; for raising funds for them internally by taxes; for preparing budgets of the resources available from taxes, from allocation of land revenue, and from matching grants by the states; and for spending the funds carrying out the projects. The responsibilities of the panchayats extend throughout the entire field of rural life—in the more social areas of water supply, sanitation, village health, primary education and educational facilities; in such economic overhead areas as the construction of rural roads and supply of power to the village; and in specific agricultural production activities such as local irrigation and channel construction, improvement of seeds, land reclamation, and distribution of

[1] I am here again trying to generalize for India, but I recognize that the variations from state to state are substantial since the laws setting up the new institutions are state laws. For a useful review, see. V. T. Krishnamachari, *Report on Indian and State Administrative Services and Problems of District Administration,* Part II (GOI Planning Commission, 1962).

fertilizers. These functions are the responsibility of the local panchayat for the individual village, and they extend to the higher level where projects overlap a single village and where problems of finance and budgeting require approval on a higher level. It is hoped that this new system will eventually contribute to rural planning from the bottom up, rather than on the basis of targets imposed from New Delhi and often not achieved, except on paper, at the local level. This system is the so-called *panchayat raj*. It has been introduced recently in answer to the criticism of the Community Development Program that the villagers were not sufficiently or deeply drawn into the development process, but its origin can also be found in the positive ideas of Jai Prakash Narayan for a change in the structure of government to devolve authority to the village through a "grass roots," village democracy free of politics and faction. This elected framework is paralleled by an existing bureaucratic hierarchy of officers who are carrying out the new functions of development. There is some combination of both of these elected and bureaucratic lines at the district level, since the district council frequently has as a member the district collector, that is, the senior district administrative officer, and it may have a small, paid secretariat.

The elected village panchayats may be free from party politics; they are not free from village politics. Their relation with village politics is somewhat subtle since many of the elected village panchayat co-opt a scheduled caste member and a woman member. Furthermore, it is desired as a matter of policy to avoid politics, and to create an illusion of the absence of politics a unanimous vote is preferred—in fact some states give grants to the village if they achieve such a vote.

It is clear that this by itself raises an entire series of problems like the relation of the traditional panchayat to the elected one; the role of caste and faction in the election; and the creation of unanimity in its possible absence. But these questions are directly related to a far more important question: Under what condition does a village panchayat function effectively for its purpose of mobilizing interest and effort on the part of the villager for development?

In fact, the effective functioning of the panchayat depends largely upon the role of caste and faction within the village and the interests of its dominant group. Where the traditional dominant caste retains power unthreatened, the elected panchayat in fact may simply be a democratic gloss on the powers of the dominant caste: the panchayat members not of the dominant caste are simply appointees of the dominant caste leaders, and the panchayat chairman is simply the traditional village head or his appointee.

Where the village is ridden by caste splits or factions, the effect may be more-or-less true elections; but then the mass of villagers are often unwilling to grant the panchayat any powers or to yield it anything since the various groups do not trust each other. This may happen if there is intercaste rivalry or factionalism within a caste reflecting changes in caste status or aspiration for a change. In such cases, there may be a legal panchayat reflecting the power of a caste group, but the refusal by a significant part of the village population to recognize the legal panchayat in effect deprives it of any power.

However, if the village consists largely of either a single caste or related castes, and if the leaders of this caste do see the advantages of certain projects, then the village panchayat may function effectively. Unanimity may in fact be achieved voluntarily—normally done by informal discussion prior to the formal election. But this success, even in a more-or-less united village, in large part depends upon the individual panchayat leader—the chairman elected by the members of the panchayat—who plays the key official role in the village. Around his election the village rivalries frequently center. In villages that have chosen a younger man with vigor, knowledge of the development program, and the confidence of the villagers in his integrity and ability, as well as the support of the older traditional leaders, the results have been reasonably successful.

Given this analysis of the possibilities for success, it is not surprising that the program on the village level has not been as immediately successful as the idealists had hoped. At the same time it has had more success than the pessimists give it credit for. Considering that until recently the villagers had no experience of any local assembly, or of the powers of the chairman; that they never previously had either taxed themselves or budgeted a program for the village; that there had been factionalism and distrust within the village; and that these programs are closely involved with factional interests—the new program even so, has had certain desirable consequences. In some villages it has led to greater trust among the village groups, even to informal coalitions. It has contributed to some consensus on projects and some willingness to raise funds to the order of several thousand rupees for the agreed-upon projects from the village. It has brought into prominence in the village a somewhat younger group of leaders than previously, has given them experience in joint action, and has made the villagers more aware of development. In these respects the new elected panchayats have been reasonably successful.

However, the panchyats in many cases have not had much effect upon agricultural output. One of the major aims of the program was to lay greater stress upon raising output and to plan for such increases, but in many villages these problems are considered to be questions for the individual peasant. Although the panchayat is able to decide on all-village projects—such as a school or a well, where these projects do not raise questions that give rise to caste or factional conflicts—they have apparently entered into local projects of a directly agricultural nature far less often.

> [The] present mentality [of the village panchayats] appears to be . . . that they should interest themselves in such matters as are of common concern to the village and which by their nature cannot be the subject of individual concern. . . . Now, obviously, schools, wells, roads, dispensaries, etc., fall in the first category; agriculture falls in the second. . . . Secondly, the Panchayat funds belong to the whole village. Any improvement of an irrigation source does not ordinarily benefit all people in the village; it benefits only some, and even to them the benefit may be in varying proportions. Even if a Panchayat is production minded, which it is ordinarily not, it cannot easily take up works of agricultural improvement in view of the above difficulty.[2]

Above the village panchayats are the block and district panchayats. The block panchayats frequently consist of the elected chairmen of the village panchayats within the block, with possibly some additional elected members; the district panchayat often consists of the chairmen of the block panchayats. Frequently too there are elections—directly by the villagers or indirectly via the block panchayats—to the membership of the district panchayats. There is some overlapping of functions between the village and block panchayats; the district panchayats have a general supervisory role, and also a crucial budget-approval role. Although party politics are hopefully absent from elections to the higher-level panchayats, they often play a definite role in elections for district panchayat members. The funds available at the district level are relatively large: in Madhya Pradesh it is estimated that about Rs 3 million are spent in a district in the areas of panchayat responsibility, and in Maharashtra the district budgets have averaged in excess of Rs 15 million, a much greater amount than before the new panchayat raj system. The control over such amounts of money is a major source of political power to the heads of the district panchayats, and its expenditure creates political obligations and support in the villages within a district.

---

[2] Association of Voluntary Agencies for Rural Development, *Report of a Study Team on Democratic Decentralization in Rajasthan* (New Delhi: February, 1961), p. 16. See also Association of Voluntary Agencies for Rural Development, *Report of a Study Team on Panchayati Raj in Andhra Pradesh* (New Delhi: October, 1961); on the general area of success see Planning Commission, Programme Evaluation Organization. *Some Successful Panchayats—Case Studies* (New Delhi: Government of India, 1960); Ralph H. Retzlaff, *Village Government in India* (Bombay: Asia Publishing House, 1962); William and Charlotte Wiser, *Behind Mud Walls, 1930–1960,* (Berkeley and Los Angeles: University of California Press, 1963), paperbound ed. chaps. x–xii on the changes.

At best a grant means both employment and improvements in a village; and at worst, it results in the enrichment of Village Committee members. In either case, the rural leaders are doing the village a favour which can be put in a credit balance of obligations, to be cashed in the form of votes on committees or in elections.[3]

On one hand the effect of this greater power makes elections to the district and block panchayats a matter of political and party importance, whether open or covert, since control over the panchayats wins blocks of votes. Another effect in Madhya Pradesh, and probably in other states, has been to increase the strength of rural political leaders within the Congress: first, to make them rivals to urban politicians, rather than mere allies or supporters of them and thereby able to bargain hard with candidates under meaningful threats of resignation from the party; and second, to challenge the position of the district member of the legislature. Formerly the member of the legislature had the major power in the district by his ability to influence the state bureaucracy on expenditure and projects within a village; but with control over rural projects and expenditures in the hands of the new panchayats, chairmanship of a block or district panchayat may be more important, in rural areas at least, than election to the legislature. At the same time, with the new rural political power, urban members of the legislature in constituencies that include rural areas become, of necessity, much more interested in rural matters and stronger supporters of rural projects than before.

From the data available on the candidates for them and the memberships of the district panchayats in Madhya Pradesh and Maharashtra, it appears that the candidates—especially in the Congress party—are landowners, and often substantial ones. In a Maharashtra district, for which a good deal of data on all candidates except the Communists are available, the candidates generally own more than five acres of land, are in the higher income brackets, are relatively well educated, and know more than just the state language. Caste has not been a major issue in this district because the Maratha caste is so numerically dominant. Many of the candidates, especially in the Congress party, are also members or officers of coöperatives and other new rural financial institutions. (Interestingly enough few are members of caste panchayats.) Shrader points out that in this district, where caste factors have not played a part, the Congress party had such control of patronage and such influence with the state

[3] A. C. Mayer, "Rural Leaders and the Indian General Election," *Asian Survey*, Vol. I, No. 8 (October, 1961); this article and the one by L. L. Shrader and R. Joshi, "Zilla Parishad Elections in Maharashtra and the District Political Elite," *Asian Survey*, III, No. 3, (March, 1963), pp. 143–156, are major sources for the politics of district panchayats.

and national governments that there was little reason to oppose it: therefore no opposition party has much chance. Furthermore, he stresses that in this relatively homogeneous district, only the middle-class and the rich peasants have the experience and breadth of interest to provide district leadership. In a district panchayat election in Gujerat, however, caste issues were strong because a rising and large group of small landholders of one caste were challenging the economically stronger, but numerically smaller, existing leadership: the latter traditionally led the Congress; the former had joined the Swatantra. The defeat of the Congress by the Swatantra in the panchayat elections was directly associated with this caste conflict.[4] Thus on the wider district level the panchayats are deeply involved in politics, and although they are to some degree apparently simply repeating aspects of the larger political party struggles and issues, they are, in fact, providing rural groups direct experience in elections and district government. They provide new openings and another forum for the play of factions and caste elements in party politics as well as a place for newer and younger elements in local politics to begin. So they create challenges to the existing legislative leadership within the Congress by bringing up a somewhat younger group of leaders with a great deal of local economic power. However, there seems to be little evidence that groups other than the landowning, dominant castes are playing an active role in these panchayat organizations, even though both scheduled castes members and, in some states, women are necessarily members by law. From the one study that has been done on this subject, the main political factors in allocation of funds within the panchayat do not appear to be narrow caste factors, but rather factors that center around efforts to build up and maintain a working political coalition.[5]

Together with the new, popular organization extending from the village to the district there is the bureaucratic organization. The Community Development Organization plays the main bureaucratic role in economic development in the rural areas. The organizational framework of the community development work consists of a project area of about 300 villages with a population of about 200,000, within this there are several development blocks of about 100 villages each and with a population of about 65,000 to 70,000 per block, and within these blocks there are about twenty groups of villages with each group being served by the village-level worker. These community development officials are under the

[4] R. Kothari and G. Shah, "Caste Orientation of Political Factions," *The Economic Weekly*, XV, Nos. 28–30 (Special Number; July, 1963), pp. 1169–1178.

[5] A. C. Mayer, "Some Political Implications of Community Development in India," *Archives of European Sociology*, Vol. IV, 1963, pp. 86–106.

Ministry of Community Development, a central ministry. There are separate central government Ministries of Food and Agriculture and of Power and Irrigation, which are also involved in these programs.

The states have responsibility for programs in health, education, agriculture, rural industries, and cooperatives—all areas overlapping the functions of the community development officers. At the district levels each of the appropriate state ministries provides technical officers in each field, and all these officers at the district level are responsible to the chief administrative officer, the district collector. Below him the block development officer of the Ministry of Community Development, who is responsible for the block community development work and reports to the collector as well as his ministry, coördinates the work of the technical officers in his block; and at the village level, the village-level worker (*gram sevak*) tries to carry out the program. With the introduction of the elected panchayats into the development field, each of the bureaucratic officers works with the panchayat at his level in terms of formulating programs and budgets and carrying them out. But the formal relationship varies from state to state. In some cases it is informal but in others the bureaucratic officer is himself a member of the higher-level panchayat. Much of the purpose of the village panchayat program was to get away from the bureaucratic curse of directives from New Delhi, which were more or less meaningless at the village level, and apparently the program is having some effect in this direction; but this takes time since it must also overcome the deep distrust of the government official on the part of the villager and the practice of verbal acquiescence to bureaucrats. It has also had the effect of making the bureaucratic official more responsive to local political influences. Thus the position of the elected panchayats in the rural development program is at this time a somewhat uneasy relationship with the bureaucracy. Attitudes with respect to this relationship range from statements that the panchayat's role is to assist in carrying out the plans formulated at the central and state levels, to statements that the central and state rural plans will be built up from the individual village programs.

The village-level worker, who is the person in this bureaucratic hierarchy in direct contact with the villagers, supposedly stimulates desired change and effects it too. In itself this is a mixed role; he must not only educate the villagers as to the desirability of a change, but, as part of the process of education, he must know as much about the change as any of the villagers. He thus must be sufficiently educated to have a broad understanding of the national program of social and economic change, he must be able to communicate effectively with the villagers, and he must have available sufficient technical knowledge to convince the peasant of the worth of a suggestion. He must also always be aware of targets and goals in specific fields that he must fulfill if he is to move ahead in the bureaucracy. This is a difficult combination of tasks, and there are relatively few young men with these capabilities. The rapid expansion of the program by the states after the initial successes in a few sample areas has meant that the demands for qualified men have exceeded the supply: this is one reason the program has bogged down.

Another important problem is the conflict between the possible social goals, which may involve mixture of castes and bringing in untouchables (quite apart from the more formal steps to reduce caste disabilities) and the production goals, which involve dealing with the landowning farmers, usually members of the dominant castes. Efforts in one area can jeopardize efforts to communicate in other areas.

In the early stages, major emphasis was placed on social change; at present, major emphasis is being placed on increasing farm output. But the latter may be at the expense of the confidence of the nonlandowning castes, whether clean or not; at the same time the worker often lacks the real technical competence that the peasant has. A new seed or technique he recommends, on suggestions from above, may be unsuitable to the area, which is manifest to an experienced peasant, who knows it will fail after a trial or after one good year. The risk in the introduction of new techniques is high; and once confidence is lost or the peasant loses money on the new crop, his faith in the village-level worker is destroyed.

Almost of necessity, the worker tends to be much closer to the members of the dominant caste than to the other villagers. These are the peasants who own the land and greatly influence the attitudes of the villagers. They are also the most potent group politically. The new worker obviously looks for the leaders when he comes into a village, and working with the existing village leadership may prove costly for his real effectiveness in fulfilling the goals of rural development.

Quite apart from this basic problem of incompatible goals, the sheer variety of tasks and of required reports in very many fields puts a great strain on the worker's time; when this is combined with the training difficulty, and with the problem of simply becoming acquainted with a new area, it is not difficult to see why the original community development program was not very successful.[6] The panchayat raj program was

[6] The literature on this program is voluminous. I recommend especially for a brief general review, United Nations, Department of Economic and Social Affairs, Commissioner for Technical Assistance, *Report of a Community Development Evaluation Mission in India, 23 November 1958–3 April 1959,* "Report No. TAO/IND/31" (August 17, 1959); S. C. Dube, *India's Changing Villages* (London: Routledge and Kegan Paul, 1958), especially Appendix I; Committee on Plan Projects, *Report of the Team for the Study of Community Projects and National*

specifically introduced to associate the village more intimately with community development, but it is still an open question whether it can overcome the similar problems.

Another related institutional change in the area of rural development was the introduction of a network of coöperative credit institutions into the villages, pyramiding into a hierarchy of credit institutions at the state and national levels. The use of coöperative credit has certainly increased, but debt financed from coöperative credit societies is still only a small proportion of the total farm debt outstanding. In 1953, it was about 4 per cent; it has been estimated at 6 per cent in 1959/1960. I have pointed out that there was a tendency for political leadership in the local Congress party and in the elected panchayat to be associated with office holding in the coöperative organization, and with such characteristics as landowning and membership in the dominant caste of the area. Control over local coöperatives is unquestionably a source of political strength and a reward for it. Certainly credit is made available far more readily to farmers who own land, and generally larger farmers are favored—on legitimate grounds and possibly others. Sir Malcolm Darling, one of the moving spirits of coöperative credit in India under the British, has estimated that only about one-third of the peasants are "credit worthy." There is also a greater tendency for the larger farmers, unlike the smaller farmers, to use agricultural credit they do receive for production rather than consumption purposes, and this has unquestionably contributed to the relatively greater growth of output by the larger farmers. It has also made it possible for the larger farmers to withhold their crops from market longer in a hope for higher prices between harvests. It is claimed too that this access to credit has improved the financial position of the larger farmers and rural moneylenders, who in turn may then lend to smaller peasants on the conventional terms.[7]

In some states coöperative sugar factories have sprung up in large numbers. These are closely associated with the dominant landowning castes in their areas, and by their power over purchase of sugar from the peasants they have great political influence. As a result, the head of the

Extension Service, "The Report of the Balvantraj G. Mehta Commission of the GOI" (3 vols.; New Delhi: November, 1959), is probably the most detailed analysis of the development in India, and this led directly to the introduction of panchayat raj.

[7] Sir Malcolm Darling, *Report on Certain Aspects of Cooperative Movement in India* (GOI Planning Commission, 1957); *Rural Credit Follow-Up Survey,* "1958–59 General Review Report" (Bombay: Reserve Bank of India, 1961). These are among the good short reviews of the situation. See also S. A. Shah, *et al.,* "Distribution of Agricultural Income in India," *The Economic Weekly,* Vol. XV, No. 41 (October 12, 1963). A more recent study is D. Thorner, *Agricultural Cooperatives in India* (Bombay: Asia Publishing House, 1964).

local sugar coöperative is very often a Congress party leader in the area, and the head of the hierarchy of coöperative units in the state has great state-wide political power in the party.

Thus the various institutions at the village level, elected and bureaucratic, contribute primarily to broadening the political-economic strength of the landholding peasants, especially of those castes with greater numbers. The shifts in political power at the village level reflect the greater role of population number and votes, but they are associated with economic power. For those castes without land or access to new sources of income, there has been little immediate improvement on the village level, and the new government institutions have not played much of a role. For the lower castes and untouchables, however, the access to education on a much wider scale than ever before, as well as the reservation of some educational and job opportunities, offers hope for future economic improvement. But where economic improvement has occurred for a caste, scheduled or not, the new institutions provide access to still further economic power and concomitant political power.

THE STATE GOVERNMENTS

The question of state government is a very neglected area of Indian research, so there is little written matter on which to present conclusions [8] in spite of the fact that the state governments have such great power under the new constitutions. In contrast to the village the state offers a far wider area of political conflict, and the subcaste elements which are so important on the village level are subsumed in broader caste and communal issues. Furthermore, policy questions become far more important than on the village level. Since the state official, elected or appointed, is dealing with general problems that may impinge on national problems, holding an elective or relatively high appointive position in the state government calls for a degree of education, experience in dealing with people and with problems, and a skill in administration that are all above the requirements on the village level.

The state reorganization of 1956 and thereafter, relating state boundaries to regional languages, has probably had the effect of increasing the role of caste factors within state governments. The limits of similar subcaste structures and similar caste status are the linguistic boundaries, and these are now often the state boundaries. In multilingual states the

[8] Probably the best of recent studies specifically related to state politics is F. G. Bailey, *Politics and Social Change*, (Berkeley and Los Angeles: University of California Press, 1963), which stresses the changes in the character of politics from the local to state levels in Orissa.

castes of one area would counteract the castes of other areas, and the whole field of potential coalition is widened; at the same time the role of any caste association is weakened, since it would probably not include castes in other language area. However, even in the new linguistic states any state is large enough so that one caste, or even a group of allied castes, requires support of other castes or factions; and in consequence the roles of factions and cliques are often very strong, both in the state parties and in the state governments. The states are also responsible for a whole range of powers and activities that impinge directly upon the peasant, so that state legislation on land reform, rural taxation, education, and road transport is of direct interest to rural caste and communal groups.

Cities and urban groups play a smaller role in state politics than in national politics. The large urban industrialists, the trade-unions, and intellectuals have national interests, for many of the decisions of interest to these groups are made by the national rather than the state governments. For the dominant rural groups the state governments play an active and positive role, whereas their interest in the national government is more passive, centering about a desire to have friendly government so that the state will get "its share." For the urban groups the situation is the reverse: their interest in the state government is relatively passive, more or less limited to a desire for a friendly government that will provide police protection for their industries and will lobby for them; but their interest in the national government is very active because the national government makes the positive decisions in the field of industry and commerce.[9] In effect the rural, more traditional forces, which play such a major role in party politics—especially in the Congress party—also play the major role in the state governments in the choice of ministers and the policies of the ministries. This has profound implications for policy in those areas in which the state governments have power, especially in such fields of agricultural policy as the willingness to proceed further with land reform and with agricultural taxation policy; and the implications are not positive for policy in the urban improvement field.

The bulk of the discussion of the role of the bureaucracy is in the section of this chapter on central government. Since what I say there is also relevant to the state government, I shall only briefly indicate some differences in the position at each level. First, the local pressures on the state bureaucracy are stronger than on the national. The Member of the Legislative Assembly, the MLA, is far more a representative of his local constituents than the M.P. in the Lok Sabha; and he is specifically expected to support his constituents in detailed dealings with the bureauc-

[9] These generalizations admittedly have many loopholes.

racy. Furthermore, the MLA is likely to be less experienced in the relationship between himself and the bureaucrat, and may therefore seek to exert pressure more directly. Caste and communal factors play a more important role in selection and promotion of officials in the state bureaucracy than at the national level. Since the reorganization of the states on a linguistic basis, the communal, linguistic factor may determine selection at the top level of the bureaucracy, for ability to speak the state language is important. On the state level also, members of the bureaucracy are more likely to be associated with particular intragovernment or ministry cliques based on communal or geographic factors than they are on the national level, since these factors are more important in the state. It is also true that the state bureaucracy is likely to be less well-paid and less well-trained than at the national level, although this is more likely to be true below the highest levels than at the highest. This, of course, affects the relative quality of the work done by the two bureaucracies.

In the economic development field, the state governments have great influence in the area of planning and in the implementation of plans, but they have in the past rarely had planning organizations. Rather, the state plans, which are forwarded to the Planning Commission, have often been a collection of projects that the Chief Minister, various operating ministries, and other political leaders thought desirable, with little analysis or choice among the projects. In the process, the leading ministers in some of the states have played a significant, direct role in the industrial development of the states, or a role through members of their families and other associates. The former Chief Minister of Orissa, B. Patnaik, moved from industry, in which many of the firms he had set up were sponsored and aided by the state government, into politics and became Chief Minister. P. S. Kairon, former Chief Minister of the Punjab was criticized by a judicial inquiry and resigned because members of his family apparently used his position and power to assist their economic schemes.[10]

However, the most important fact about the state governments is that they have, over the past decade, become much more powerful in relation to the Center. The reasons for this are not hard to see. First, they have many responsibilities in key areas of government policy and action, especially in rural affairs, which directly affect the great majority of the Indian people and their incomes. Thus, the states are very important political centers of power. Second, the Congress party organization repre-

---

[10] See on Orissa the series of four articles by G. S. Bhargava in *The Indian Express,* July 23, 24, 25 and 27, 1964; on the Punjab see articles in *The Times of India,* June 14 and 15, 1964, as well as numerous earlier articles in that newspaper, *The Economic Weekly,* and other journals.

sents the play of the more traditional forces in Indian life, especially in rural India; and the party organization had been far more powerful in influencing policy at the state than at the national level, for it had been relatively weak in relation to Prime Minister Nehru. Third, the states especially since the reorganization in 1956, represent the strong regional and linguistic elements in Indian life, and these forces exert a good deal of pressure in India independent of the centralizing forces. The state governments speak for their inhabitants in bargaining for increased funds and new projects, and the competition is not only between inhabitants of two states—say, of the type between California and New York for defense contracts—but between the Bengali community and the Gujerati- or Hindi-speaking communities. The chief minister of a state, if he controls or works harmoniously with his state Congress party organization and is considered to represent the people of his state, will be a very powerful political force; at the Center, a minister is under the shadow of the Prime Minister and in fierce competition with other cabinet ministers. Under the circumstances the attractiveness of the Center for the ablest regional politicians has greatly diminished, especially when it can mean a loss of state power to their successors in state cabinets; and as their experience with the potential powers of the state governments increases, they find that the states permit large play to their abilities. Thus, the longer-run factors that make state governments more powerful also contribute to attracting abler political figures to the states; in turn the Center attracts either a less able political figure, or one with limited chances for advancement on the state level—because of caste or other factors.

THE CENTRAL GOVERNMENT

THE PRIME MINISTER AND HIS CABINET

India had for almost all of its recent existence as an independent nation the same Prime Minister, Jawaharlal Nehru. From the death of Sardar Patel in 1950 until his own death in 1964, Nehru was not only head of the government but the unrivaled, even if not always the official, head of the national Congress party. His position was unique, and it is most unlikely that, at least in the reasonably near future, his successors will exercise the power that he exercised at his peak or be able to use the methods that he used. At the same time, his ideas and methods will continue to have a profound effect upon party and national policies. Some examination of his role is of great importance in understanding the functioning of the government.

First, Nehru was a national figure, with a position stemming from his role in the Independence struggle and his relation to Gandhi. He was vitally associated with the Congress party and played a major role in its internal conflicts and its elections, but he was above the party. Nehru represented the nation; his appeal was far beyond the party as such, which was the reason for his great strength within it when he did use it, and his decisions were made from a national viewpoint. This national viewpoint was further encouraged by his own major interest in international affairs—he was also Minister for Foreign Affairs—in which it was necessary to look at the national interest as a whole. His ideas and attitudes, representative in many ways of the urban, intellectual, English-educated upper class of India and strongly influenced by the Fabian socialism of England in the 1930's profoundly affected the national policies adopted by the Congress movement. His ideas are now part of India's laws and policies: the secular democratic state, the nonaligned foreign policy, economic planning stressing large-scale industrialization, a socialistic pattern of society, the subconscious distrust of business and the belief in administrative controls, and the need for social change. At the same time his pragmatic approach, his willingness to compromise, his personal alliances, his administrative quality, his belief in his personal role—all set the tone of the government and determined the implementation of the policies. Meanwhile his capacity to appeal to the great mass of the population created support for the party and its policies and provided India with the leadership that made so much of the difference between its history since Independence and those of its neighbors, such as Indonesia and Pakistan.

Apart from his position as Minister of Foreign Affairs, Prime Minister Nehru was also Chairman of the Planning Commission, which he was mainly responsible for establishing, and he took an active personal role in contributing to the Plans. His support was basic to the prestige and power the Commission has in fact had. Therefore his view of the planning and development process is of great interest:

> [Planning] and development have become a sort of mathematical problem which may be worked out scientifically. . . . [Planning] for industrial development is generally accepted as a matter of mathematical formula. . . . [Men] of science, planners, experts, who approach our problems from purely a scientific point of view [rather than an ideological one] . . . agree, broadly, that given certain pre-conditions of development, industrialization and all that, certain exact conclusions follow almost as a matter of course.[11]

---

[11] R. K. Karanjia, *The Mind of Mr. Nehru* (London: George Allen & Unwin, Ltd., 1960), pp. 49–52.

Accepting this notion of the semiautomatic characteristics of planning and development, plus a lack of real interest in economics as compared with foreign or political problems, he inevitably played only a *deus ex machina* role in economic problems. Once the Plan had been written, and in this writing he normally contributed heavily to the introductory and general chapters, he played only a minor role in implementing it or in resolving its problems. He was interested in achieving the broad goals of "development" and "socialism," and also strongly favored "coöperative farming." But he was much less interested in defining these policies or their implementation; as a result some of these policies were stillborn, and conflicts in the implementation of the plans or the policies often remained unresolved among the competing ministries.

Apart from being the administrative head of certain specific ministries, Nehru was also the head of the Cabinet. Within the Cabinet, until his illness, the Prime Minister was indisputably the chief. The other members were appointed by him; and although he recognized personal, policy, regional, and communal groupings in these appointments, and although at times of great pressure from the party he had to yield to it—as in the dismissal of Krishna Menon from the Defense Ministry—his was still the dominant position. The other ministers were below him in power and range of responsibilities. They were also far more limited in terms of their responsibilities for specific ministries, as representatives of specific regions of communities, and for various groups within the Congress. The Prime Minister always attempted to maintain something of a balance within the cabinet, including representatives of the so-called right and left wings within the party, such minority groups as the Moslems and the scheduled castes. The policy orientations of the ministers were very loosely defined, and they in fact often defied their classification. For example, the so-called right-wing Finance Minister, Morarji Desai, introduced a severe defense budget with such innovations as a compulsory deposit scheme, and also began a vigorous effort to curb gold hoarding; his successor, T. T. Krishnamachari, supposedly less conservative, abandoned the deposit scheme and relaxed the gold control program. But they were expected to be somewhat representative of their geographic areas and communal groups; they are known as much, if not more, for the state or community they come from as for their national portfolios. Members have resigned where the interests of their areas have been unduly neglected—as in the example of the former Finance Minister, C. D. Deshmukh, over the failure to create a separate state of Maharashtra. The mixture of such ethnic and policy interests also prevented caste and communal factors from playing a major role at the Center. The various ethnic and regional pressures cancelled each other out on the general Cabinet level if not within the individual ministries, especially with Prime Minister Nehru's known attitude.

Nehru was generally able to surmount conflicts among the ministers and force a united policy by his own prestige and party leadership, at least until his first illnesses of 1962. Thereafter the conflicts among the ministers combined with the increasing struggle over the succession, and the failure of past policies with respect to China resulted in greatly weakening the functioning of the Cabinet.

The Prime Ministers and the members of the Cabinet, apart from their political position, are also the heads of the bureaucratic ministries for which they have political responsibility. To a certain degree Prime Minister Nehru, in his period of health, became involved in many of the details in all ministries, when they interested him.

This involvement throughout the whole government was Nehru's own style. It reflected his lack of interest in administrative neatness, but it also reflected his view of himself as not only the administrative head but the personal head of government. As such, he was fulfilling the traditional role of a ruler as the court of final appeal for the Indian people as a whole. But this involvement also made clear the Prime Minister's picture of himself as head of the Indian bureaucracy. India was fortunate in inheriting a group of trained and able administrators. Much of the success India has achieved since Independence lies in the ideological agreements these administrators shared with Nehru and the Congress party program, and in the resultant effectiveness of the administrators in formulating and carrying out policies to achieve those programs. At the same time the solutions adopted on many of the problems reflect weaknesses within the bureaucratic structure.

THE BUREAUCRACY AS A POLITICAL FORCE

Within the government at the central, state, and local levels, the bureaucracy is not only an administrative agency but a political force. The administrators do make politically important decisions, either in the course of administering broad policy or in preparing questions for the minister's decision. But the role of the bureaucrat has changed since Independence.

Under British rule until the war, the government's role was relatively minor, and the administrator made decisions; in fact appeal over his head was difficult and was more or less confined within the bureaucratic hierarchy. This has changed in several respects. The government is far more actively involved in economic life than ever before, so that there are many more economic decisions to make than ever before—decisions for preparing a five year plan, running a government-owned steel plant or machine tool factory, or carrying out a village development program.

On the other hand, all of these decisions can now be appealed, not only to a higher rank within the bureaucracy, but outside it to political leaders from the local level to the national level. Although the type of personal decision making of the old district officer or the local ruler has diminished, the range of appeal on decisions has become wider and the power of the individual bureaucrat or ruler has become more circumscribed.

One of the most obvious signs of the increased scope of bureaucratic activity is the great increase in the number of government officials at all levels. This increase in numbers is itself of great political importance, for these positions are to some extent patronage for the Congress party or ruling factions within it. At the same time the officers are a new channel between the village and the government, and thus play a political role within the village.

In central and state government policy the role of two overlapping groups within the bureaucracy are crucial: the top civil servants within the government ministries or institutions, who are still in many areas members of the old Indian Civil Service (ICS) or, if not, of the new Indian Administration Service (IAS), and the officials of the Ministry of Finance, who must in effect approve the actual expenditure of all items within the budget, and who control the foreign exchange applications. Both groups of officials in effect do make the decisions that carry out the policies, and they often lay out the lines of policy, in the absence of decisions by the political leaders, that are subject to appeal to the leaders. The Indian top bureaucracy is as well trained for decision making as any equivalent group in any country of the world. There are some problems, however. First, these officials are generalists and trained as generalists. As such they are considered able to perform well in any government job of suitable rank, and there is a policy of fairly frequent rotation. This approach may be reasonable with respect to top policy questions, but its value with respect to industrial operations is much more dubious. To run an industrial enterprise, or a railroad system, or the coal board takes time and experience: frequent rotation on such jobs cannot add to efficiency. Second, these officials are trained primarily in making the day-to-day decisions that constantly arise in government. This is a most important characteristic of a good administrator, and under the British it was undoubtedly the most important one, since the British interest in general changes in the economy or society was slight. Today such decision making is only one part of government activities, for there is added now a need for foresight, discernment of relationships within the economy and among policies, and planning capacity. These skills, as distinct from the willingness and capacity to make decisions speedily, are rarer among the top echelons of the Indian administration. Combined with overwork and constant pressure of decisions, this lack contributes to a failure to foresee problems that may be imminent; it has contributed to tendencies to look at issues from a narrow agency point of view and to operate within traditional administrative techniques and controls.[12] This approach by these officials may be supported by such factors as training or attitudes toward their own role.

I have the impression that the power in India of the top government officials is very great. There is little devolution of authority. This reflects an attitude toward authority and a style of operation that was traditional within India and then reinforced by British rule. Combined with the pervasive role of the government, it contributes to a concentration of authority, to a hesitancy in making decisions and extended delays at the lower levels, and to the great importance of contacts with the right top person at the same time that the burden on the few persons at the top is extremely heavy. As a result the top secretaries of a ministry or department are extremely powerful and equally overworked, whereas the lower levels may not have enough to do.

Apart from this problem I think it fair to say that there are the normal rivalries and cliquishness between ICS and non-ICS members—not to mention within the ICS. There are others between those at the central and those at the state level; and between those appointed in operating posts in industries, with their careers in the enterprises, and those appointed from the administrative services, with their careers in the services. Similar rivalries of course exist in any large bureaucracy in any country although they may take a different form.

Almost all government economic activities funnel through the Finance Ministry since it must approve all expenditures, and this explains both the key role of this ministry and the criticism of its power. This power arises from British procedures to control expenditures, which were introduced to check and double check the handling of money by Indians. These procedures have been largely continued by the Indians, and they contribute to delay at many levels of government—especially, for obvious reasons, in the development field. They can create extremely difficult problems in carrying out a project or running an industry or an industrial enterprise, since the man in ostensible charge shares power with the finance officer. C. Subramaniam, former Minister of Steel and Heavy Industries, the ministry responsible for operating the government-owned steel and fertilizer plants, in introducing several management reforms in those plants stated:

> Historically this [finance] officer had often been regarded as having veto powers over the General Manager. . . . An aggressive Financial Adviser (represent-

[12] I am not saying these same tendencies do not exist in other governments for other reasons. I do believe they exist in the Indian government.

ing the Finance Ministry) could find occasion to intervene in almost anything. . . . If the managers of our public sector plants were mostly incompetent or corrupt, as sometimes seems to be assumed, and therefore must be hedged in with restrictions, then our public sector enterprises would progress only haltingly and at prohibitive costs. If industrialisation was to progress, then they must trust their . . . managers and give them full authority. If they betrayed that trust or failed for any cause, then they must be promptly removed.[13]

This power of the Finance Ministry combined with its regular budget power and its control of foreign exchange makes it the most powerful economic agency, and the secretaries in that ministry are the most powerful in the administrative hierarchy. Apart from Finance, several of the operating ministries, because of their age, the experience of their personnel, and the large numbers of their employees, have very great political influence, and their policies in many respects are free of control. The major ones in this class are the Railroads Ministry, which has in effect a dominant influence on transportation policy as a whole, and the Irrigation Ministry.

In the economic field in particular, there are the problems of coördination between the Planning Commission, which is an advisory body, and the Finance Ministry and the operating ministries that carry out the plan. These are not problems peculiar to India; they exist in most countries with Planning Commissions. A recent newspaper article referred to the fact.

Some failures of the Plans are partly due to the commission's inability to keep a proper eye on what goes on in the Central Ministries. The most palpable instance is the foreign exchange crises of 1958 where, if the commission had intervened in time and halted the licensing policy pursued by the Commerce and Industry Ministry, the Second Plan would have been saved its biggest trial.[14]

This crisis also reflected a lack of coördination between the Commerce and Industry Ministry and the Finance Ministry, with the result that awareness of the drain was lacking and action was not taken until it was almost too late. Similar problems have arisen in the field of transport, power, and coal; and in 1963 there were public disagreements and criticism over agricultural policies between G. L. Nanda, then Minister for Planning, and S. K. Patil, then Minister for Food and Agriculture. The Planning Commission, which prepares the Plan and has little or no coördinating power, depends largely for its facts upon the operating ministries. The role of the economists within the government is greatest in the Planning Commission, and is far smaller in the operating ministries.

---

[13] "Public Sector Plants," *The Hindu Weekly Review*, September 30, 1963, p. 7.
[14] *Times of India*, May 16, 1963.

This unquestionably contributes to the fact that the plans prepared and the advice given by the Planning Commission economists may have little influence upon the actual economic policies of the ministries implementing the plans.

Finally, the administrators themselves are both individuals and members of caste and class groups with individual or group interests, and these interests may influence the policies they present for action—including the plans and laws they write—and the actions they themselves take. It has been frequently pointed out that the land-reform laws are in many cases administered by officers who are themselves landowners and the caste legislation by people themselves of high caste. The administrators interested in urging or pushing the allocation of funds and projects for the benefit of certain states come from those states; they may also be influenced by the possibility of future employment, for themselves or members of their own families, and this is obviously the reason for the public worry about the retirement of these officials to private industry.[15] I think it would be fair to say that although such factors play a role in decision making by the bureaucracy in India, they do so less than in many other countries; there is a tradition of objectivity, and the bureaucracy, including the military, has not been an overt political force. In the Delhi area itself the Jan Sangh, a Hindu communal party, is quite strong, and it is believed that it derives a substantial part of its strength from lower middle-class government officials of relatively high caste status. However, the officials supporting the Jan Sangh are not policy makers, and there is little connection between their politics and government policy.

These are elements within the administrative system itself that contribute to the character of economic policy, if it is defined broadly to include both the planning and administering on such policy. There are other elements that center about the relation of the administrators with the political leaders, the ministers and the Lok Sabha. British rule was a version of a benevolent despotism and the administrators were able to operate within a wide area of freedom from political limitations, with a good deal of independence and a guarantee of support from their superiors.

This situation is inevitably different in a self-ruled democracy of the Indian type. The administrators are under a minister, who is fully responsible for the ministry, and who is an elected political figure; thus the administrators are directly subject to internal political pressures. Those

---

[15] In recent years the problem of corruption has been stressed in the press, and even by ministers. It is expected that this problem would increase as both the demand for permits has arisen, and the discrepancy between government and private incomes has increased. Without playing down the importance of this I suspect there is significant exaggeration of the rumors, especially at the national level. In comparison with many other countries, corruption in India is relatively low, although standards are relatively high.

pressures are influenced both by the stakes at issue, which are now much greater than under the British because of the great government involvement in economic life, and by the variety of political forces within the Congress party and its relations to the opposition. In India this inevitably means that the whole gamut of caste, communal, class, and regional pressures that exist within the party also exert pressure on the administrator. The administrator must educate and persuade the political leader of a policy, and he is subject to an appeal to the political leader. Furthermore, the minister not only represents the nation, but he is also frequently in the Cabinet as a representative of an area, a group, or of a faction within the party whose political future lies in an area or a group. This has the effect that he may try to staff his ministry at the higher levels with officials from the same or similar groups whom he can trust—which, of course, happens in other countries as well—and these officials look upon other ministries as competitors for power or for scarce resources. Contact among the ministries may therefore be slight, with each one functioning to some degree as a more or less independent satrapy. If a strong prime minister or the prime minister working together with other ministries provides effective leadership on policy for the whole Cabinet, these conflicts may be of minor importance and the independence of the satrapies will be in fact small; but if this leadership is lacking the conflicts among ministries may be serious and the results, administrative chaos.

The major requirement for administrative decision making is mutual confidence between the minister and his top civil servants. Where this is lacking, the administrator will avoid making decisions or delay them in an effort to refer everything to the minister. Sardar Patel, by his support of the ICS, encouraged this mutual confidence. The effect of the Life Insurance case [16] upon the morale of the administrators was poor, and from all reports it had the effects, at least in the short run, both of discouraging independent responsibility and initiative on their part and of encouraging red tape such as written minutes.

Ministers will, of course, vary in their willingness to delegate responsibilities to their subordinate ministers or their civil servants. Sardar Patel is still spoken of as an able administrator who did delegate; however, administration was not considered one of Prime Minister Nehru's strong points. Prime Minister Shastri was supposedly better in this respect. Dr. Roy, the former Chief Minister of West Bengal, delegated very little, bypassing his ministers and civil servants regularly. The state government was notorious for its delays, and everything moved at the interest and speed of Dr. Roy. His great ability and decisiveness prevented complete

[16] W. H. Morris-Jones, *The Government and Politics of India,* pp. 116, 134; also T. Zinkin, *Reporting India* (London: Chatto and Windus, 1962).

chaos, and even resulted in movement. One of the first moves of his successor, P. C. Sen, was to require that all civil servants report through their ministers; and in consequence the role of the ministers has become much more important, and the operation of the administration has become more orderly than under Dr. Roy. This may, to some extent, make up for the lesser strength of his successor. If a minister is weak or not interested in his ministry, this in effect gives the senior administrators the power, up to a point, since someone must make the decisions. However, the senior administrators do not have independent political power, so their influence and scope of action is severely limited, and the ministry will very probably be a weak one. In the economic sphere the members of the bureaucracy have been of great importance in making policy. Although political factors are very important in economic policy, they are of less direct importance than in other aspects of government. This means that many of the problems are considered technical problems to be handled on a bureaucratic level. Prime Minister Nehru himself had little interest in the details of economic policy, so that his political influence was not involved in economic policy. Many of the ministers, in ministries that were dealing with economic matters even if they were not directly economic ministries, have only a slight knowledge of economics and are willing to leave the economic problems to their secretaries and undersecretaries. Thus for all of these reasons the role of the bureaucracy in the operating ministry is especially strong in making economic policy. It is certainly stronger than that of the professional economists, whose bureaucratic position is not high, and who are largely confined to the advisory Planning Commission. It may be stronger than the political leaders on many issues where the political implications are not particularly obvious.

The role of the Lok Sabha and the state assemblies in this relationship is complex. The parliamentary bodies pass the budgets, approve the plans, and set up the government-owned industrial and commercial enterprises. The ministers, the agencies, and the officials running government enterprises and disbursing government funds are of course ultimately responsible to the Lok Sabha for the Center and the assemblies for the states; at the Center both the Estimates Committee of the Lok Sabha and the Auditor General of India are essentially investigators reporting either to the Lok Sabha or to the president, independent of the Cabinet.

The members of the Lok Sabha can and do raise questions on the most minute as well as general points of the operation of the ministries and the enterprises, and these questions are raised not only by opposition members but by Congress members. It was a Congress member who brought up the Life Insurance case by questioning in the Lok Sabha. However, while Lok Sabha members seem to lack any custom of limiting questions to general ones, the ministers normally do not report to Lok Sabha committees, nor keep interested members, operating through informal committees, regularly informed on the activities of their ministries. Thus these public questions are the main method of eliciting information for the individual member. This means that many questions of rather technical detail, in terms of allocation of funds or detailed decisions to do one thing rather than another, can come up for questioning in the Lok Sabha well after the event, and the responsible officer may find himself having to justify relatively minor actions taken years back. For a timid bureaucrat, the fear of these post mortems can inhibit any action and encourage prolonged buck-passing, especially if the minister's willingness to support him is questionable or if the party is plagued by factions and the ministers frequently change.

It was the purpose of this section to point out the role of the bureaucracy as a political force in development. The bureaucracy on the whole favors greater power to the Center; the government officials at the top are national officials, even though they may be deputed to state or industrial positions. They are members of a group whose function is to wield power, and they are trained in its exercise. In the case of a weak minister, they probably make policy; even with a strong minister they must decide the many administrative problems and details that create the broader problems, and the alternative choices placed before any minister for a broader decision are often prepared by the administrators and thus influenced by them. They are inevitably far more closely tied in with democratic Indian politics than they were under the British. They are no longer platonic guardians; but they are subject to political ministers representing not only the nation, but states, classes and castes, and they influence ministers with a political future in India. They are subject to political criticism and pressures both from the Congress and its opposition, as well as from conflict with other ministries and other factions within their own ministry. In India the administrative manner and policies are still largely set by the ICS; but this small group also has its style, its limitations, and its point of view that encourage an ad hoc approach and administrative controls. This has a profound effect on Indian economic policy, planning, and administration. The training and attitudes of the ICS administrators on the whole agreed with those of Prime Minister Nehru, and this undoubtedly contributed to the past successes of the economic effort; at the same time the attitudes of the ICS contribute to at least some of the present problems of planning.

RELATIONS BETWEEN THE CENTER AND THE STATES

The Center's powers that give it strength in dealing with the states are: First, that the nation is India, and although the heads of the state parties and governments are very powerful within their boundaries, they have neither the prestige nor the power of the prime minister. Second, the Center is responsible for planning, for industrial policy, for trade and foreign exchange, and for allocating resources among projects and among states. Third, although the states obviously have their own powers of taxation and revenue-raising, they also depend in large part upon allocation of taxes from the Center for state funds; this allocation is carried out by an impartial body under the Constitution, but it also makes the states dependent upon the Center. Fourth, state laws must be approved by the president, which in the Indian system means that the prime minister and the Cabinet may raise objections and approval may not be granted; and if the President decides that the government of a state is in danger of breaking down, he may supersede it for a period of time. Also, the central government has emergency powers which it may use under declared circumstances to override the states; and it obviously has powers of defense. Finally, a comparison of the central and state governments would show that most of the state Congress parties are riven with factionalism far more than the national party—in part because the factors contributing to factionalism operate almost unchecked on the state level—and that the quality of the central bureaucracy is probably better than the quality of that in the states, especially at levels below the very top. The Center has great powers, and the role of the resurgent state governments and parties is not so much a positive forming of national economic policy, but rather a negative attempt to get greater allocation of revenues or taxes or a greater number of projects to one, rather than to another, state.

With the increasing strength of state leaders within the Congress party, and with the greater state consciousness since the states now consist largely of people of the same linguistic groups, the pressures the states can exert upon the Center are strong; and this has a major effect upon the allocation of projects and resources among the states in the Plans. Some of this is inevitable, but I think the relationship between the Center and states combines centrifugal and centripetal forces. On the achievement of Independence, with the leadership at the Center, the obviously national shape of the life and death problems, the ideological bias toward central policy, and the *élan* of Independence and a successful national revolution—it was not surprising that the centripetal forces were greatest. But as the country solved the immediate problems; as many of the national leaders died and were not replaced, or tired and became older; as the traditional rural forces and groups within the country became stronger, especially at the state level once the linguistic states were created, and as they achieved greater power in the Congress party—it is not surprising that centrifugal forces have become greater. These latter forces are unlikely to become so great that India will split; the states and the leadership of the states have too much to gain by remaining united, both in terms of actual resources and in terms of an Indian nationalism.[17]

It is very improbable that the forces at the Center—national, bureaucratic, military and political—would permit a split voluntarily; and the demand for this on the part of the states, except in Madras where the DMK at one time wanted a separate Tamilnad, has not been politically important. The strength of the national feeling shown during the brief Chinese and Pakistan incidents was also very heartening. Except in the event of a major war, or a military defeat or establishment of a dictatorship, the centripetal forces should also not become overwhelming. Rather, a balance may be expected in which the states retain much of their present powers to influence central decisions. With the greater urgency of defense since the Chinese attack, with consequently larger expenditures for defense by the central government, and with the successors of Nehru being younger and possibly better administrators—the Center has the potential to restore some of its influence, although central power is unlikely to reach near its past peak without an emergency or a drastic political change.

The clearest example of the greater power of the state leaders was the choice of Nehru's successor, made largely by what appears to be an effective coalition of the stronger regional political leaders, the most important of whom K. Kamaraj, was and is the head of the national Congress party. He was the first party head since the death of Sardar Patel who had sources of power independent of Prime Minister Nehru. This is the most obvious example of the increased power of the state leaders, especially when there is harmony between the state government and the party. Even here what should be stressed is their power of choice. A prime minister, because of his powers, may become a stronger figure than Nehru was in 1962–1964, although unlikely to achieve the strength of Nehru at his peak.

SUMMARY: THE POLITICAL CHANGES SINCE INDEPENDENCE

There has been a shift in power within the Congress from urban to rural groups. This shift in power has been most obviously reflected in the government by the changing character of the legislatures.

Within rural society since the introduction of universal suffrage, num-

[17] S. Harrison, *India: The Most Dangerous Decade,* has stressed the influence of these centrifugal forces especially. B. R. Nayar, in his "Contemporary Political Leadership in the Punjab" (Doctoral thesis, University of Chicago, 1963) stresses rightly, I believe, the countervailing centripetal forces, especially both the central powers and Congress party functioning, which have contributed to unity in India.

bers have become a much greater factor in determining caste dominance. Rural castes that are both numerically strong and have sufficient economic position and awareness to be affected by the economic policies of the government have sought to improve their political position, and in many cases they have done so. These castes are largely in the class of landowners with medium-size holdings. The effect of this and of the new panchayat raj system has been to disperse power more widely within the Congress.

Urban classes that have wealth and are vitally interested in the economic policies of the government have also increased their political power by serving as sources of party finance; their influence is especially exercised on administrative policies. In urban areas also community factors have become more important as a source of members for votes.

The intellectuals, agitators, and social reformers who had such strong influence in the Congress movement have lost power. The main strength of the intellectuals within the party was former Prime Minister Nehru and, within the government, the higher members of the bureaucracy. However, the economic goal of development, first urged only by the intellectuals and a relatively small urban group, has over the decade become much more widely accepted by the Indian population. I think it is legitimate now to talk of "rising expectations" which a government must fulfill, even though there is disagreement over the policies to achieve that goal.

The bureaucracy at all levels, from the village to the Center, has become far more deeply involved in politics than heretofore. Government officials have lost the range of independent action they once had, while their influence is far more pervasive with the great expansion of government activity in India. With this spread in activity the scope of power of the bureaucracy is great, especially in carrying out existing policies and preparing alternatives ones.

The state parties and governments have greatly increased their strength vis-à-vis the central government since Independence. The balance of power may even have swung to the states. At the same time there are historical, political, and economic factors that would favor some strengthening of the power of the Center.

PART IV

Economic Policy and Achievement, 1947–1962

39. The State shall, in particular, direct its policy toward securing:

(a) that the citizens, men and women equally, have the right to an adequate means of livelihood;

(b) that the ownership and control of the material resources of the community are so distributed as best to subserve the common good;

(c) that the operation of the economic system does not result in the concentration of wealth and means of production to the common detriment; . . .

41. The State shall, within the limits of its economic capacity and development, make effective provision for securing the right to work.

43. The State shall endeavor to secure, by suitable legislation or economic organization or in any other way, to all workers, agricultural, industrial or otherwise, work, a living wage, conditions of work ensuring a decent standard of life and full enjoyment of leisure and social and cultural opportunities.

46. The State shall promote with special care the educational and economic interests of the weaker sections of the people.

The Constitution of India,
Directive Principles of
State Policy, 1950

6

Economic Programs: Ideologies, Strategies, Achievements

INTRODUCTION

The economic ideas of the various groups that formed the Congress movement reflected the urban strength of the movement—the urban or westernized intellectuals and the industrialists who favored large-scale industrialization. But they also reflected the Gandhians, who presented a program of local development centering about village industry and handicrafts. Although theirs was a rural program, the followers of Gandhi were rarely peasants, but rather social reformers seeking to improve the lot of the peasants.

The urban intellectuals in the first group tended to favor a future socialist India, at this stage ill-defined, but including at minimum a large amount of government control and increased government ownership of industry. However, Nehru laid stress upon a democratic transition to socialism, which was to be achieved in coöperation with the private sector and not at its expense. In the agricultural field they laid emphasis on land reform and wanted eventually a coöperative system of agriculture. The urban industrialists, who favored large-scale industrialization, were neutral if not antagonistic to this vision of a new socialist India; and they were relatively indifferent to agricultural policy. Although the Gandhians were deeply interested in agricultural policy and, in general, were in favor of the ideas of equality and social reform contained in socialism, they feared the centralization that both socialism and large-scale industry imply and did not accept the class conflict basis of Marxian socialism. They agreed that if any movement toward socialism were to occur, it should be gradual without harming any group in society.

To the extent that the peasants were interested in policy questions, they were involved in questions of land reform, rural coöperation, and the like, but were relatively indifferent to industrialization and socialism; and their position was not uniform. Since the princes and the largest landowners, who were often intermediaries or absentee landlords, were often anti-Congress, their influence on Congress policy was slight. The smaller but still medium-sized landholders and cultivators often paid tribute to these intermediaries; title to their land was frequently at the mercy of the large landholders. Still, they played a positive role within the Congress, especially as supporters of Gandhi; they favored land reform, but their enthusiasm for coöperative farming or landholding and for social reforms that attacked the caste system was not great. The landless peasants and untouchables had little interest in or influence upon policy; they were more often the object of interest, rather than the centers of power. When they participated in making policy, they tended to favor policies that would improve their social and economic positions although some of these policies might be opposed by other important peasant groups. Finally, the interest of the peasants was not so much in long-deferred benefits for either themselves or India, for they were first interested in avoiding any worsening of their own economic position—even if it might mean a stronger India at some future date—and, at the same time, they were interested in anything that would raise their incomes in the near future.

The change in the Congress party from a struggling coalition, in which urban intellectuals played a large role, to a governing coalition, in which their importance declined sharply while that of rural and urban business groups rose, was indicated by economic policies reflecting the attitudes of the newly powerful groups. However, this relationship was not abrupt or very obvious in India. Nehru, who was one of the leaders of the intellectuals, was head of both the national government and party still. Also, on many individual parts of the programs advocated by different groups, it was possible to get sufficient agreement from the different elements within the Congress to carry those parts into law. Finally, the style of Indian politics is one in which there may be unanimity on a public statement of policy but not on its implementation, so that public plans or statements might follow rather than precede shifts in policy.

In India, the Five Year Plans present economic goals and programs for an extended period. By examining plan goals and plan achievements, it should be possible to see which sectors have benefited most from the Plans. The Plans require policies to carry them out, and those policies are end products of political struggles among groups within Congress. At least on some issues, the decisions on programs and projects within the Plans are related to the described shift in political power; the detailed policies adopted for the implementation and administration of the Plans are influenced by such factors as the growing power of the states and the changing powers of the identified groups; and finally, to some extent at least, those groups that have gained in political power also gained in terms of such measures of economic well-being as income. At the same time there cannot be a 100 per cent relationship between political gain and income gain, for shifts in the economy apart from political factors influence income distribution and gains among various groups—as, for example, the farmers in the United States, who though strong politically, have lost relatively in income terms. But one would expect some association among these variables, especially in India where the government's role is so pervasive within the economy; and it may be possible to measure income shifts among these groups.

Part IV will examine the ideology and strategy of the Plans and their achievements, the decisions on specific policies to implement the strategies, and the distribution of the gains arising from the progress made. From these it will be possible to show more clearly the relationship between the political and social changes since Independence, on one hand, and the economic changes on the other.

IDEOLOGY

The Congress party came into power with a general program of economic development. A major ideological force behind this program was Prime Minister Nehru, and it was a program that was developed, in its broad terms, prior to Independence. In 1938, Congress had already set up a planning committee headed by Nehru to look at the economic problems of the country and prepare a program of policies to deal with them if power were achieved. The ideological basis of that program can be expressed best in Nehru's own words written before Independence:

> Inevitably we are led to the only possible solution—the establishment of a socialist order, first within national boundaries . . . with a controlled production and distribution of wealth for the public good. . . . [This] can hardly take place without the willing consent or acquiescence of the great majority of the people concerned.
>
> Is it desirable or possible for us to stop the functioning of big-scale machinery in our country? . . . It is obvious that we cannot do so. If we have railways, bridges, transport facilities, etc., we must produce them ourselves or depend on others. If we want to have the means of defense we must not only have the basic industries but a highly developed industrial system. No country today is really independent or capable of resisting aggression unless it is industrially developed.[1]
>
> The cooperative principle should be applied to the exploitation of land' by developing collective and cooperative farms. It was not proposed, however, to rule out peasant farming in small holdings . . . but no intermediaries of the type of talukdars, zamindars, etc., should be recognized after the transition period was over. . . . [Banks], insurance, etc. . . . should at least be under the control of the state, thus leading to a state regulation of capital and credit. It was also desirable to control the export and import trade.

[1] Jawaharlal Nehru, *An Autobiography* (London: Bodley Head, 1953 [first published in 1936]), pp. 523, 526.

These programs were to be achieved through planning. The outline of the Plan being worked on before Independence,

> . . . was inevitably leading us towards establishing some of the fundamentals of the socialist structure. It was limiting the acquisitive factor in society, removing many of the barriers to growth, and thus leading to a rapidly expanding social structure. It was based on planning for the benefit of the common man, raising his standards greatly, giving him opportunities of growth, and releasing an enormous amount of latent talent and capacity. And all this was to be attempted in the context of democratic freedom and with a large measure of cooperation of some at least of the groups who were normally opposed to socialistic doctrine. That cooperation seemed to me worthwhile even if it involved toning down or weakening the plan in some respects.[2]

Stress must be laid upon the basic elements of democracy and coöperation in this vision, and its clear implication that the achievement of this coöperation was not to be sacrificed to ideology.

Nehru's general point of view represented that of many of the intellectuals in the Congress, but it was also sympathetic to many intellectuals and officials not in the Congress. Some of the specific aspects of Nehru's thought were viewed favorably by leading industrialists, who prepared their own plan during the war. Gandhi's point of view, with its emphasis upon village industry and the handicrafts, was contrary in philosophy to programs of centralized planning and large-scale industry; but the Gandhians did not have an influential role in the government after Independence. They did, however, support the idea of democratic change.

PLANS

Independence brought a host of political and economic problems that had to be solved immediately, so the preparation of a general plan was necessarily shelved. The Congress party in 1947 set up an economic program committee headed by the prime minister; in 1948 this committee recommended that a permanent Planning Commission be established. This was in fact set up in 1950, again with the prime minister as its chairman. The responsibility of this Planning Commission to prepare a Five Year Plan for India was the equivalent in the economic sphere of the drafting of the new constitution in the political sphere. As the new constitution signified the end of the political transition after Independence, so the new Planning Commission symbolized the transition from a day-to-day solution of the economic problems associated with the postwar tran-

[2] Jawaharlal Nehru, *The Discovery of India* (London: Meridian Books, Ltd., 1956 [first published in 1946]), pp. 405–406.

sition to a conscious effort to achieve the desired goals of economic development in India.

The First Five Year Plan, which appeared shortly before the first national elections, was a beginning document issued as much for its psychological and political effects by showing action toward future, stated goals by the Congress party, as for its economic consequences. The Plan itself was based largely on projects and programs that were either already under way or in the files, and the overall goals and targets were a final addition to, rather than a point of departure for, detailed planning. This was unavoidable under the circumstances. There were good monsoons in the last two years of the Plan, which stimulated agricultural output, farm prices, farm demand, real wages, and profits. In combination with the government commitment to the Plan and the increased expenditures themselves, the existing, idle industrial plant was put to use again to nearly full capacity and new investment was begun. The Plan was completed "successfully": output and income rose to the planned levels, prices fell somewhat compared with the 1950/1951 level, foreign exchange reserves were drawn on only slightly, and—perhaps most important—it was felt that India could step up its pace of development markedly during the Second Plan.

With the Second Plan, which was presented in 1956, comprehensive planning really began. An income target was the starting point. A model of the development process was presented, and provided the basis of the Plan. There was an extended period of research and preparation, and an integrated Plan was in fact prepared. The assumptions, goals, and strategy of the Second Plan also provided the basis for the Third Plan, which was published in 1961.

The literature on planning in India is now voluminous. Apart from the Plans themselves—including such official documents as "The Second Plan Framework," and various official preliminary research papers, as well as a large unofficial literature available in India—three recent books published in the United States and readily available to non-Indian readers present much of the data of the Plans and their results, review much of the literature, and present critical evaluations.[3]

The main goals of Indian planning are: (1) to increase per capita income, primarily by raising total output (relatively little stress has been

---

[3] On the early history of planning in India see P. C. Mahalanobis, "Heralding a New Epoch," *Talks on Planning* (Calcutta: Indian Statistical Institute, 1961). For the best brief official presentation of the function of the Planning Commission and the planning procedures in India, see GOI Planning Commission, *The Planning Process* (Delhi: 1963). The three recent books on Indian planning published in the United States are: J. P. Lewis, *Quiet Crisis in India* (Brookings, 1962); W. Malenbaum, *Prospects for Indian Development* (Free Press, 1962); and W. B. Reddaway, *The Development of the Indian Economy* (Irwin, 1962).

placed on the population denominator);[4] (2) to place Indian growth on a self-sustaining basis that is no longer so heavily dependent on fluctuations in agricultural output, which in turn depend so greatly on the vagaries of the annual monsoon; (3) to make the Indian economy and its development less dependent on a few raw materials and traditional exports, which are subject to world market price fluctuations; and, at the same time, to make India eventually independent of foreign aid, which would contribute in turn to its freedom in foreign policy and its political independence; (4) in this process to provide increased employment for the unemployed and underemployed, as well as for the large numbers annually entering the labor force; and (5) less immediately, to diminish the inequalities in income and status among persons, groups, and regions.

The strategy to achieve these goals lays major stress upon the development of heavy industries.

> In the long run, the rate of industrialization and the growth of national economy would depend on the increasing production of coal, electricity, iron and steel, heavy machinery, heavy chemicals, and the heavy industries generally which would increase the capacity for capital formation. One important aim is to make India independent, as quickly as possible, of foreign imports of producer goods so that the accumulation of capital would not be hampered by difficulties in securing supplies of essential producer goods from other countries. The heavy industries must, therefore, be expanded with all possible speed.[5]

Associated with this development was the expansion of such capital-intensive economic overhead sectors as electric power and railroad transportation.

While the Second Plan was being prepared, the lack of foreign exchange was not such an overriding constraint on the economic plans as it was subsequently; but the increasing stringency in foreign exchange, which developed during the Plan, reinforced in practice the planners' emphasis upon the expansion of the capital-goods sector of the economy, with major stress laid upon import-substitution.[6] It was recognized that the development of the capital-goods industries would make only a minor direct contribution to the unemployment problem. At the same time, investment in these industries would create increasing monetary demand in the short

[4] Compared with other underdeveloped countries, India's public stress on population control is relatively high.

[5] GOI Planning Commission, *Papers Relating to the Formulation of the Second Five-Year Plan* (New Delhi: 1955).

[6] For an ex post theoretical justification of this see: K. N. Raj and A. K. Sen, "Alternative Patterns of Growth Under Conditions of Stagnant Export Earnings," *Oxford Economic Papers,* XIII, No. 1 (February, 1961), pp. 43–52.

run, which, although not increasing the supply of consumer goods, would thus contribute to rising consumer prices. For both these reasons it was hoped to expand production in the labor-intensive household and handicraft industries, while restricting output of the factory-produced consumer goods where they were competitive.

Agricultural output, too, would have to be increased to supply foodstuffs to meet the greater demands, because of higher money incomes arising from the Plans, from the growing population and its increasing industrialization and urbanization. The construction of irrigation works and the like in rural areas would contribute to solving the short-run employment problem. This increase in agricultural output was also to be stimulated by a program of land reform that would not only create incentives for greater farm output, but would contribute to the social goals of greater equality of income and wealth. Finally, various social programs toward such goals as widening education and improving health were to be pursued both to raise the productivity of the Indian laborer and to create a sense of improved well-being on the part of the population.

The effect of this program upon the Indian economy as a whole would hopefully be profound. Although specific goals were changed in the light of the achievements, the Third Five Year Plan indicated these goals for the fifteen years from 1950/1951 to 1965/1966: an increase in national income of close to 90 per cent; an increase in per capita income of about 35 per cent; and an increase in the level of investment and savings from approximately 5–8 per cent of India's national income to approximately 15 per cent for investment and 12 per cent for savings.

With the pressure of population upon the land and with the importance given to industry in the Plans, it was hoped that from 1950/1951 to 1960/1961 the percentage of the labor force employed in agriculture and allied pursuits would fall from approximately 72 per cent to 68 per cent. At the same time the contribution of agriculture and allied pursuits to national income would decline from 49 per cent to 46 per cent.

The Third Five Year Plan, published in 1961, briefly discussed longer perspectives. The hope was expressed that national income would approximately triple from 1950/1951 to 1975/1976, per capita income would rise by close to 90 per cent, the percentage of the national income invested and saved would rise to about 19 per cent, and the proportion of the labor force engaged in agricultural pursuits would decline to about 60 per cent. This last repeats a hope in the Second Plan.[7]

---

[7] Of the three paragraphs above describing targets and hopes, the first and third are from chaps. ii and x in GOI Planning Commission, *Third Five-Year Plan* (New Delhi: 1961); the second paragraph is from chap. iv of the "Draft Plan Frame" of

The main instrument for the expansion of the capital-goods industries, as well as the main channel for increasing investment and savings, was to be the Government of India—defined to include both central and state governments. The government alone had the power to raise the resources necessary for a greatly expanded program of investment which was to be concentrated in the heavy industry sector. But it was not only the scope of the Plans that led to the much greater role of the government; the ideology of planning in India, based upon the ideal of a socialist society, led logically to a major role for the government. This was formalized by a resolution of the Congress party accepting a "socialistic pattern of society" as its goal and by various industrial policy statements. In these latter the industries of India were divided into three groups—the first, where future development was left to the government, either independently or jointly with the private sector; the second, where the government would take the initiative but existing private firms would be allowed to develop; and the third, which included all other industries, where all was left to private enterprise. The first two groups included the direct military industries, the major economic-overhead industries, and the basic capital-goods industries—ordnance, rail and air transport, shipbuilding, generation and transmission of electric power, mining of many minerals including coal, and oil drilling. They included production of basic iron and steel, heavy ferrous castings and forgings, machine tools, heavy plant and machinery for basic industries and power plants, basic ferroalloy and nonferrous metals including aluminum, and various basic, intermediate, and final chemical products including antibiotics and fertilizers. These were "the commanding heights" of the new industry and economy in India; government ownership and control in these fields would both satisfy the socialist aspirations and provide a tool for the accumulation of capital for their development. However, the lists of industries were not rigid with respect to the private-public mix, and there were sufficient qualifications to permit either public or private investment in any group, depending on the circumstances.

To achieve this major increase in government investment in the capital-goods sector it would be necessary at the early stage of the development process for the government to divert savings, foreign exchange, and other resources from the household and private sectors of the economy to its own use by taxation, direct allocation, and so on. This would be inevitable since at the start of the development process the government's own contribution to national income and to savings, through the budget, was

---

the *Second Five-Year Plan,* as reprinted in P. C. Mahalanobis, "Heralding a New Epoch," pp. 39–40. See also pp. 12–15 in GOI Planning Commission, *Second Five-Year Plan* (New Delhi: 1956).

minor. As development proceeded, however, it was hoped that the contribution by the government sector to the supply of capital would increase as a result of the earnings of the new government-owned industries.

It was recognized that both the concentration of investment in the capital-goods sector and the increased demand for and diversion of resources this required would create bottlenecks and shortages. If so, increased government intervention in the form of steps to ration scarce resources and to control their prices would prove necessary.

The role of the Planning Commission in this entire planning process is to prepare the Plans, which would coördinate the various investment proposals from the government and private sectors into a total national program and indicate the required resources to carry out the plans. In fact, the commission's detailed planning effort on the investment side has been largely confined to government projects, but the larger private projects had to be individually approved and included within the Plan before they could proceed. On the resource side, the commission's role has been more indicative of the domestic and foreign sums required, with the Finance Ministry having the responsibility of raising and allocating financial resources. The commission also prepares periodic evaluations of progress under the Plans and has an advisory role in dealing with problems as they arise. The actual administration of the planned programs is left to the operating ministries concerned in the central government and to the state governments in those areas, especially agriculture, in which their responsibility lies. The coördination of the work of the commission with the ministries and states is formally insured by the fact that the prime minister and the central ministers most deeply involved in economic problems are members of the commission; likewise, the chief ministers of the states are members of the National Development Council, which must approve the plans prepared by the Planning Commission. As indicated earlier, the problem of coördinating the activities of the commission and the operating agencies is in fact one of the serious planning problems.

Without entering into the plans in detail, the best summary of their emphasis is derived by looking at the public outlays and their sectoral allocation during the first two Five Year Plans and in the Third Plan, as shown in Table 9. The Third Plan called for a very great stepping-up of the rate of government expenditure with a total for five years exceeding the previous ten-year total. The breakdown in Table 9 is by functional sector. Unfortunately, it is not possible to break down expenditures between urban and rural sectors with any exactness. However, if it is estimated that all the agricultural outlay; that conservatively half of the outlays for education, rehabilitation, and health and all the outlays for "welfare of backward classes;" that half of the outlays for "road and road transport" and roughly one-fourth of those for railways and for communications; and finally that half of the noninventory, miscellaneous outlays are in the rural sector—then approximately 45 per cent of the actual public outlays in the Second Plan were in the rural sector, and 40 per cent of the proposed outlays in the Third Plan will have been.

Although in the First Plan public-sector investment was just about 49 per cent of the total, in the Second Plan it increased to 55 per cent, and in the Third Plan it is to reach 60 per cent, which reflects the fact that almost 70 per cent of the total additional investment in the Third Plan was allocated to the public sector. This increased spending in the public sector was to be heavily concentrated in expansion of the steel industry primarily, but also in electric power, nonferrous metal, machine building and machine tools, fertilizer and other basic chemical industries.[8]

GENERAL ACHIEVEMENTS, 1951–1961

The main immediate purpose of the first two Plans was to accelerate the pace of economic development in India. To contrast it with the pace of the preceding periods, the economist must have a measure; these changes can best be measured by income and wealth. Very little is known precisely about either the level or the trend of income in India from 1800 to 1945. Obviously total income rose or India would not have been able to support its much greater population, but whether the rate of increase in income exceeded or fell below the rate of increase in population is not known, and opinions on the matter are diametrically opposed.[9] There is greater agreement that the period between the end of the First World War and 1950 was one of accelerated population growth, and that the increase in total income probably lagged behind the rate of population growth so that per capita income did not increase, and may even have fallen. One of India's leading economists, Professor V. K. R. V. Rao, has described India in the period from 1931 to 1951 as "a static economy in progress." [10] This term may be considered at best a convenient shorthand summary of the development in those two decades, which saw the rise of new industries, a war boom, and a postwar recession.

[8] W. Malenbaum, *Prospects for Indian Development* (New York: The Free Press of Glencoe, Illinois, 1962), p. 248; *The Third Plan,* p. 59. These investment estimates in the Plan are very crude and approximate; it is also not defined whether they are gross or net figures.

[9] For the most recent argument that per capita income fell, see V. V. Bhatt, *Economic Change and Policy in India* (Bombay: Allied Publishers, 1963); for a very convincing contrary argument, see Morris D. Morris, "Toward a Reinterpretation of 19th Century Indian Economic History," *Journal of Economic History,* XXIII, No. 4 (December, 1963), pp. 606–618 .

[10] V. K. R. V. Rao, "A Static Economy in Progress," *Capital,* December 16, 1954.

TABLE 9
SECTORAL ALLOCATION OF PUBLIC OUTLAYS IN THE FIRST THREE PLANS[a]

Sector	Actual Outlays in First Two Plans: Rupees (billions)		Actual Outlays in First Two Plans: Per Cent	Planned Outlays in Third Plan: Rupees (billions)		Planned Outlays in Third Plan: Per Cent
Agriculture		18.9	29		19.8	27
Agricultural production	4.8			}		
Community development coöperation	3.4			} 10.7		
Major and medium irrigation	8.5			6.5		
Village and small industries[b]	2.2			2.6		
Industries, mining, and power		15.4	24		25.3	33
Organized industry and minerals	9.5			15.2		
Power	5.9			10.1		
Transportation and communication		18.2	28		14.9	20
Railways	11.2			8.9		
Roads and road transport	3.8			3.0		
Other transport and shipping	2.2			2.2		
Communications	1.0			0.8		
Social services		11.4	17		12.2	16
Education	4.0			5.0		
Health	3.1			3.4		
Welfare of backward classes	1.1			1.1		
Housing, social welfare	1.5			2.3		
Rehabilitation	1.6			0.4		
Miscellaneous		1.6	2		2.8[c]	4
Totals[d]		65.6	100		75.0	100

Notes:

[a] "Outlays" includes "expenditure on the creation of physical assets . . . including expenditure on personnel required for putting up these assets," as well as non-investment expenditure on staff salaries, subsidies, and so on. It is thus more inclusive than investment. It is estimated that non-investment public outlays in the Second Five Year Plan were Rs 9.5 billion, and in the Third Plan would be Rs 12.0 billion.

[b] Some of the outlays in this subsector would not be in agriculture.

[c] Includes Rs 2.0 billion for "inventory accumulation" in the Third Plan.

[d] The figures have been rounded; differences between the total and the subtotals are due to this rounding.

SOURCE:
Third Five-Year Plan, Appendix B, pp. 738–739. See also pp. 577, 651, 681, 701, and 728.

This situation changed, and sharply with respect to income, during the 1951–1961 period. Even with the questionable data,[11] it is possible to discern major trends.

The total output, officially measured by net national output at factor cost in 1948/1949 prices, rose from Rs 88.5 billion in 1950/1951 (the first year of the First Plan) to Rs 127.5 billion in 1960/1961 (the last year of the Second Plan), an increase of 44 per cent over the decade at a compounded rate of 3.6 per cent per year. Real per capita net output over the same period rose by about 18 per cent at a compounded rate of 1.6 per cent per year.[12] Within these decade totals, the real national output from agriculture and related activities increased by 36 per cent; from industry, including small-scale producers, and mining by 43 per cent; and from commerce, transportation, and services by 57 per cent.

Agricultural output is the major component of national income and the major source of employment in India. The index number of agricultural output rose from 100 in 1949 to 139 in 1960/1961, while the index number for food grains rose to 135. As a result the per capita net domestic

[11] Since I agree fully with I. M. D. Little on the quality of the Indian statistics, I can do no better than quote his evaluation that appeared in a generally sympathetic article written in 1960, "The Strategy of Indian Development," *National Institute Economic Review*, IX (May, 1960), p. 20. "It is impossible to let Indian figures speak for themselves. Sometimes they do not exist. Often they are years out of date. Always far more background is required for their interpretation than can be given in an article. National income figures exist but are very unreliable. [One Indian economist has seriously suggested that they not be used.] There is no series of investment figures, even for the public sector. [Since 1960 several series have been constructed but they differ sharply among themselves both in total and in detail.] Several different agricultural output estimates exist, and differ widely. Accurate output estimates exist for some particular industrial products. The balance of payments estimates are good, but trade figures are very unreliable for imports. In short . . . where figures are given or used [the reader] must take them as orders of magnitude, except where the context implies otherwise, and he must not always expect sources or complete tables, which, if presented, would often shed more darkness than light." For a more detailed evaluation of Indian income statistics, see K. N. Raj, "Some Features of the Economic Growth of the Last Decade in India," and A. Rudra, "National Income Estimates—Why Not Discontinue Them?" both in *The Economic Weekly*, XIII, Nos. 4–6 (February, 1961), pp. 253–271 and 209–213.

[12] The 18% figure of the rise in per capita income is derived by the formula:

% Growth in Per Capita Income = % Growth in Aggregate Income − % Growth in Population/1 + the % Growth in Population. For the source of this formula see A. K. Biswas and M. G. Mueller, "Population Growth and Economic Development in India," *Indian Economic Journal*, Vol. 2 (January, 1955), pp. 238-246.

availability of food grains increased from 13.5 to 16.2 ounces per day from 1951 to 1961, a rise of 17 per cent. The compounded annual rate of increase of agricultural production during the period is estimated at 2.8 per cent, and of food grains at 2.5 per cent. About 60 per cent of the increase is explained by the increase in area under cultivation, and 40 per cent by a higher yield per acre.[13]

Very little data are available on changes in the proportions of the agricultural output marketed during this decade. The total production of nonfood grains (cash crops) rose by 47 per cent compared with 1949/1950, and that is faster than the production of food grains, which rose by 35 per cent. This would have the effect of raising the proportion of marketed crop to total crop; the increased urbanization would have a similar effect. However, the effects of land reforms and ceilings, to the degree that they reduced the size of holdings, would probably reduce marketed amounts slightly; the large absolute increases in rural population and employment would also. The general increased well-being of the peasants, as well as the reduced impact of land taxes (see Chapter 8), may have somewhat reduced the pressures to sell also. On the institutional side, the capacity of the peasant to hold his output from market for longer periods has apparently increased. Paradoxically this has been the consequence of government and coöperative credit programs, which, by making cash resources available, has made it easier for the larger and more credit-worthy farmers to hold produce off the market for longer periods in the hope of higher prices. Although from year to year the rates of changes of the price indexes of agricultural commodities and of nonagricultural commodities were different, the overall relationship between the two sets of prices showed little change in the decade of the first two Five Year Plans. Both indexes were at about 126 or 127 in 1961 (1952/1953 = 100). Although the terms of trade deteriorated with respect to iron and steel manufactures and cement, they had improved with respect to fertilizers and kerosene, and showed little deterioration or gain with respect to leather goods and cotton manufactures. Thus on all counts there is little reason to believe that the proportions marketed have changed greatly;

[13] These figures are from the GOI Ministry of Food and Agriculture, *Economic Survey of Indian Agriculture, 1960–61* (New Delhi: 1962), especially chapter ii, pp. 6, 21–23. See also Appendix C on the extent of comparability of the statistics over time. D. and A. Thorner, "Elusive Agricultural Output Figures," *Land and Labour in India* (Bombay: Asia Publishing House, 1962), is a cautionary article on the comparability of absolute figures and the index numbers, and K. N. Raj, "Some Features of Economic Growth of the Last Decade in India," cautions with respect to agricultural income data and cites other figures.

the absolute amounts going to market would have increased with the greater output.[14]

To look in another field, the index of industrial production, including mining, manufacturing, and electric power, rose from 73.0 in 1951 to 139.0 in 1961 (1956 = 100); within this, the index of manufacturing rose from 72.3 to 136.3 and that of electricity from 60.9 to 198.8. The increases in the overall index and in the index of manufacturing output were about 90 per cent, while that for electric power was over 200 per cent. The compounded annual rate of increase in industrial output as a whole was thus about 6.5 per cent per year in this period;[15] for manufacturing it was about the same, while for electric power—whose weight in the index is very small—it was 12.2 per cent.

As a final measure of the overall economic changes within the Indian economy, and granting the rough character of many of the statistics, I present in Table 10 the Reserve Bank's estimate of the change in the tangible wealth, i.e., India's capital stock, both including and excluding the value of land, from 1949/1950 to 1960/1961. There was a 50 per cent increase in total tangible wealth measured in current prices during the two Plan periods. Since the increase in the value of land simply reflects the increase in agricultural prices during the period, it is excluded; therefore, current value of reproducible wealth rose by 90 to 100 per cent. The combined price index of machinery, transport equipment, and construction costs rose by approximately 25 per cent, thus the total real value of reproducible tangible wealth increased by approximately 65 to 75 per cent. As a result of this greater capital stock, during the period from 1949/1950 to 1960/1961, there was an increase in the ratio of reproducible tangible wealth to net domestic output from 1.8:1 to 2.3:1 (expressed in current prices). This would reflect the planned emphasis on investment in the more capital-intensive and longer-gestating economic-overhead facilities and capital-goods industries.

I estimate conservatively that approximately one-third of this increase was a product of direct government investment, or about Rs 50 billion out of the total increase of Rs 150 billion. Although in 1949/1950 the household sector had 71 per cent of the tangible wealth and the shares of

---

[14] The data in the above paragraphs are from *Economic Survey of Agriculture, 1960–61*, chaps. ii and iii; and GOI Ministry of Food and Agriculture, *Report on Market Arrivals of Foodgrains, 1958–59 Season* (New Delhi: 1959), especially the summary, pp. 1–14.

[15] GOI Central Statistical Organization. *Monthly Statistics of the Production of Selected Industries of India for October, 1962* (Calcutta: 1962), especially the Introduction (pp. iii–xii), the Note on the Construction of the Index, and the Index itself (pp. 137–171). The base was changed to 1956 to allow for changes in the mix during the Plans.

the organized business and government sectors were 12 and 17 per cent respectively, in 1961 the share of the household sector had declined to 60 per cent and the shares of the organized business sector and government sector had risen to 15 per cent and 25 per cent respectively.[16]

TABLE 10
ESTIMATE OF TANGIBLE WEALTH IN INDIA [a]
(million rupees)

Sector	1949–1950		1960–1961	
Agriculture, animal husbandry, and allied activity	52,360		87,830	
Agricultural implements		3,630		8,600
Livestock		24,280		27,020
Sheds, barns, etc.		8,800		13,660
Improvement of land (including irrigation): private		13,040		24,150
public		2,290		13,650
Other activities (excluding tea plantations)		320		750
Mining and manufacturing	25,790		68,420	
Mining		1,100		1,830
Electric generation and transmission (private and public)		2,400		11,740
Factory establishments: private		12,060		32,360
public		1,270		7,670
Tea plantations		1,330		2,820
Small enterprises		7,630		12,000
Transport and communications	23,870		42,180	
Railways		15,740		27,460
Communications and airways		1,100		2,400
Other		7,030		12,320
Government administration [b]	7,070		18,100	
Trade and commerce	17,720		33,980	
House property	44,050		71,130	
Urban: private		26,440		43,810
Rural: private		17,610		27,320
Total	170,860		321,640	
Value of land	178,540		202,410	
Total tangible wealth	349,400		524,050	

Notes:
[a] Estimated at depreciated replacement cost expressed in current prices for each year.
[b] Includes roads and bridges, capital in ports, docks, etc., airdromes, etc., and public house property.
SOURCE:
"Estimate of Tangible Wealth in India," *Reserve Bank of India Bulletin* (January, 1963), pp. 8–19.

[16] I include in this estimate of government investment in this period the increase in public land improvement (including irrigation), in public factory enterprises, in government administration, and in the government-owned railways and communication and airways activities. I also include about 80 per cent of the net investment in electric power facilities (see *The Third Plan*, p. 398). I exclude any invest-

Thus in conclusion, the first two Plan periods were extended periods of rising income and wealth, both in total and per capita terms. The rate of growth was markedly accelerated in comparison with the experience of the prior twenty to thirty years, and it probably reversed a falling trend of real per capita incomes during that earlier period.[17] This improvement reflected government policies and actions and the effect of these upon private expectations and actions. In both total and per capita income for the decade the improvement was somewhat less than they initially hoped for. This was partly because there was a far more rapid increase in population than originally projected for the decade, and it was partly because there were unanticipated lags in output, especially in agriculture. Before trying to estimate the distribution of the gains or losses experienced during the decade by the various identified groups, I shall examine the trends in the structure of employment over this period.

SECTORAL ACHIEVEMENTS, 1951–1961

One of the main goals of the Plans was to start a relative shift of resources from the agricultural sector into nonagricultural activity. It was hoped that this would begin to be slowly apparent during the decade of the 1950's. It was realized that with the growth of the Indian population and the labor force there would be little direct transfer of labor out of agri-

---

ment in small enterprises, in trade and commerce (including banks), in vehicles, transport animals, in shipping and navigation companies, as well as in other agriculture and mining. The percentage figures above are from GOI Planning Commission, "The Report of the Committee on Distribution of Incomes and Levels of Living," Part I (Delhi: 1964), p. 20. (Hereinafter cited as *Mahalanobis Committee.*)

[17] It is not my purpose to compare in detail Indian rates of growth in this decade with that achieved by other countries in comparable periods of their history. However, it may be of some interest to indicate comparative rates of growth in order to place the Indian achievement in perspective. As pointed out earlier in the text, the Indian increases in total and per capita real national income from 1951 to 1961 were respectively 44 per cent and 20 per cent. The total figures for India exceed the percentage increases per decade in total real gross national product for many of the European countries during the early stages of their development, including Russia before 1913; and they are only slightly below that achieved by Canada and Japan, and somewhat lower than the U.S. figure. Significantly, however, the rates of population growth for all the countries except the United States during comparable periods was much less than the Indian rate, so that there was much less difference in comparable increases in real per capita income for the European countries; and the Indian experience was significantly worse than that of the United States, Canada, and Japan. When compared with Latin American countries, the Indian record in both respects for the decade was much better than that of Argentina, Brazil, and Chile for the period 1925–1954; it was almost as good as that achieved in Colombia and other Latin American countries; but it was well below that of Mexico. Simon Kuznets, "Six Lectures on Economic Growth" (mimeographed, 1958), Tables 2 and 4.

culture *in toto*, but it was hoped that a relatively large proportion of the increase in the labor force would move into nonagricultural activities. Given the great pressure of population upon the land resources to start with, the probably low marginal productivity of labor in agriculture, and the apparent rate of increase in population, the direction of such a shift in the structure of the economy must be considered correct. That this aim is desirable is supported by the historical experience of many other countries that have sought to develop economically. There has been a general experience indicating that economic development is associated with the movement of population out of agriculture into industry and other employment. It is obvious too that a growing importance of the industrial and tertiary sectors would be associated with a shift in the structure of capital stock and income in favor of nonagricultural activities.

The publication of the summary volume of the 1961 Census makes it possible to examine what has happened to the labor force and population during this decade of the first two Plans.[18] From 1951 to 1961 India's total population rose by 21.5 per cent, from 361.1 to 439.2 million. This was a significantly higher rate of increase than that recorded in any previous decade since 1901, and it was much higher than the Plans anticipated. During this same decade the total population at work rose by 34 per cent, from 139.5 to 188.5 million.

Table 11 presents data with respect to the distribution of all workers in the various industrial categories of the 1961 Census for both 1951 (estimated) and 1961.[19] Table 12 presents a breakdown of the distribution of the increased number of workers from 1951 to 1961 among the various industrial categories. Both these tables show little sign of any overall shift in the structure of employment among the agriculture, industry, and tertiary sectors, with 70 per cent of the additional workers in the labor force being absorbed in agriculture.

With respect to the structure of the capital stock of the country, Table

[18] All the population and employment data in this section are from Registrar General of India, "Final Population Totals," *Census of India, 1961*, Paper No. 1 (1962). Where no specific reference is given, this is to be considered as the source. There have been changes for the better in the classification of the labor force between the 1951 and the 1961 Censuses. Therefore, the figures for employment are not directly comparable and adjustments have been made by the office of the Registrar General to make the comparison.

[19] With respect to male workers only, for whom statistics are more consistent over the ten years, the proportion of workers in the primary sector (excluding mining) fell slightly from 69 per cent to 68 per cent, increased from 12 per cent to 13 per cent in the secondary sector (including mining), and remained at 19 per cent in the tertiary sector.

TABLE 11

DISTRIBUTION OF WORKERS, 1951–1961

Industrial Category	1951[a] Millions		1951[a] Per Cent[b]		1961 Millions		1961 Per Cent[b]	
Primary sector	101.4		72		136.2		72	
Farm cultivator		69.8		50		99.5		53
Agricultural laborer		27.5		20		31.5		17
In mining, livestock, forestry		4.1[c]		3		5.2		3
Secondary sector	14.1		11		22.1		12	
Household industry		} 12.6		9		12.0		6
Manufacturing, other than household						8.0		4
Construction		1.5		1		2.1		1
Tertiary sector	24.0		17		30.2		16	
Trade and commerce		7.3		5		7.6		4
Transport, storage, communications		2.1		2		3.0		2
Other services		14.6		11		19.6		10
Total workers	139.5		100		188.5		100	
Total non-workers	217.4				249.9			
Total population[d]	356.9[e]				438.3[e]			

Notes:

[a] 1951 figures exclude Jammu and Kashmir.

[b] Differences in the two columns due to rounding.

[c] Of these, 0.8 million (or 0.6 per cent) were in mining and quarrying.

[d] It should be noted that the proportion of workers to the total population has fallen from about 61 per cent in 1951 to 57 per cent in 1961.

[e] Differences in totals in this table compared with the paragraph in the text reflect differences in coverage not reconciled in the base tables from the source.

SOURCE:

Registrar General of India, "Final Population Totals," *Census of India, 1961,* Paper No. 1 (1962), Appendix I.

TABLE 12
INCREASE IN WORKERS BY INDUSTRIAL CATEGORIES, 1951–1961

Occupational Group		Percentage Increase in Number of Workers	Proportion of Increase of All Indian Workers
Primary	1	41	60
	2	14	8
	3	25	2
Secondary	4, 5	58	15
	6	39	1
Tertiary	7	4	1
	8	40	2
	9	32	10
All workers		34 (average)	100 [a]

Note:
[a] Difference is due to rounding.
SOURCE:
Registrar General of India, "Final Population Totals," *Census of India, 1961,* Paper No. 1 (1962), pp. xxi and xxiii.

10 showed that at the start of the development process approximately 40 per cent of the country's reproducible tangible wealth (including private rural home property) was in the agricultural sector, and in 1960/1961 the proportion was about the same. This is consistent with my previous estimates that about 40 to 45 per cent of the public outlays in the Plan periods was in the rural sector, with about 30 per cent in agriculture per se; both the employment and capital structure data indicate that the character of the investment during the decade was not such as to result in a shift in the sectoral structure of the economy.

The discussion of sectoral changes in income will be left primarily to the following chapters, but it may be noted here that the official figures indicate that during the decade the estimated share of the agricultural sector in national income declined slightly from 49 per cent to 46 per cent; the share of "mining and manufacturing" remained approximately constant at 17 per cent; and the share of "services"—excluding commerce, transport, and communications which remained about constant—rose from 16 to 18 per cent. All these changes are probably well within the statistical margins of errors for the sectoral data. Although this sectoral shift in income from agriculture was desired, it is more indicative of the lag in increasing agricultural output than of fulfilled plans in all sectors.

This relatively constant structure of employment, capital stock, and income conceals a change in the structure of factory industry. Unfortunately, the published income or value-added figures are not broken down in sufficient detail over a long enough period to show the change over the decade. But the index of industrial production, constructed by weighting the components of value added, indicates that from 1951 to 1956—when the weights were changed—the contribution of textiles to the total value added by factory establishments fell from 48 per cent to 42 per cent. With employment as a measure, in 1951 the textile and the food, beverage, and tobacco industries contributed 60 per cent of the total factory employment, and the metal and machinery industries only 17 per cent; in 1959 their respective contributions were about 50 per cent and 23 per cent.

In the Second Plan, less than 10 per cent of the large-scale industrial investment was in the traditional industries of textiles and sugar; approximately 70 per cent was in the metallurgical, engineering, and chemical industries, of which almost 50 per cent alone was in the iron and steel industry. The delayed entry of the new steel capacity into full-scale operation in the Third Plan period will by itself represent an increase of over 15 per cent in the national income contributed by factory enterprises. With the delayed achievement of planned goals and the low employment consequences of this capital-goods investment, the effect of this on the employment structure of the economy as a whole, rather than for the factory-industry sector, has been minor, at least in the short run. Hopes have not been fulfilled for the start of a sectoral shift.

7

Politics and Policies: Implementation of Plans

Apart from their reflection in the Plans themselves, the policies to carry out the Plans reflect the shift in political power between rural and urban groups within the Congress since Independence. Although the stated ideology and broad strategy of Indian development plans tend primarily to reflect the influence of the urban groups, sharper political conflicts among the groups arise over the implementing policies. There are areas in which differences may be expected between politically potent urban groups—intellectuals, businessmen, government officials, and factory workers—and potent rural groups—peasants with various sizes of land-holdings, traders, and moneylenders. These occur in the agricultural policy questions like land reforms and extension of new institutions to the land. Other areas immediately affecting the position of the peasant are policy on agricultural taxes and price policy on farm products and purchases. Here also differences may be expected. Rural groups are not very much interested in the industrial field, but there are differences among urban groups over the role of the government in industry and among the states on industrial location policy. Assuming that the policies adopted are influenced by the political strengths of these groups, the resolution of these conflicts should reflect the variations in their strength within the Congress.

FARM POLICY ISSUES AND CONFLICTS

LAND REFORM AND COÖPERATIVES

Land reform is one of the major issues on which the Congress had acted both before and after Independence. This is also an area of obvious conflict between urban intellectuals and rural groups, and within the rural groups too. For the urban intellectuals land reform has been viewed as one of the main goals of social reform and a necessary preliminary for raising output on the land. For these supporters, land reform includes many policies: the first step was the abolition of intermediaries between the peasant and the state in those areas where they existed.[1] This was to

[1] The varieties of status and legal positions of these intermediaries in different Indian states were both legion and highly complex. In most areas where they

be followed by reforms concerning land tenure, size of holdings, ceilings on rent, and improved conditions for farm labor. Ultimately—this, many including former Prime Minister Nehru favored—a voluntary system of coöperative farming was to be established in order to overcome the weaknesses supposedly associated with farming small plots.

The peasants in rural areas were not united on the desirability of land reforms, and certainly not on their extent. Those who were the intermediaries would obviously suffer directly by the loss of their rights, although the amount of the loss would vary with the importance of those rights. Those who had rented land to tenants were not in favor of reducing rent; those who hired farm labor were not in favor of directly raising the wages of their laborers.

At a very early stage, even before the First Plan, various state legislatures introduced legislation for the abolition of some of the intermediaries between the peasant and state. In many cases the larger intermediaries were absentee landlords, considered to perform no economic function, but there were few of these. Prior to Independence these larger intermediaries had often been pro-British or pro-prince, and most often they were anti-Congress. They were considered "feudal relics," and the Congress party had long promised their abolition. Thus on social grounds, on the economic grounds of reducing the burden of the peasant, on the political grounds of keeping long-made promises and of removing pressure from Socialist and Communist demands for reform, and finally because of the political vulnerability of the groups affected, the state ministries moved to introduce legislation to abolish or change the position of the intermediaries. Although the legislation varied, in most states the rights of the intermediaries were abolished in exchange for compensation—partly in cash but largely in bonds—computed usually in terms of a diminishing ratio either to the former income derived from the land or to the amount of land revenue paid. However, the intermediaries received the right to resume ownership of land to which they had given up revenue rights, up to certain acreage, most often upon proof of personal cultivation. Previous cultivators of the land, if they had operated it for a long period, were often given direct holding rights under the state in whose hands ultimate rights were now vested; and shorter-term cultivators were given other rights in exchange for capital payments, which were given either in lump sum or spread out over time, to compensate

existed these intermediaries were originally given rights to collect land taxes for the state in exchange for services rendered. Under the British in some states these rights were changed to include ownership rights. In other states there were no intermediaries, and the landlord or tenant paid taxes directly.

the former intermediary. Furthermore, from the date of the legislation all payments of revenue were to be made directly to the state by the landholder. Initially the total level of payments made by the cultivator would probably not be reduced, but once the intermediary was fully compensated for the abolition of his rights, the rental payments to the state would diminish accordingly. There were obviously serious political problems for a democratic administration in preparing this legislation: such questions as the amount of compensation, method of payment, resumption rights of former intermediaries, and the rights of nonintermediary cultivators all had to be hammered out, balanced, and adjusted within the constitutionally protected property rights. Apart from the intricate complexity of the legal position of the intermediaries and its variation from state to state, the fact that there were very many small zamindars as well as a few large ones made the political problems of abolition and compensation difficult.[2] The speed of legislation, the adjudication of legal problems in the courts, and the speed of implementation of the laws also varied from state to state; but by the end of the Second Plan the legal position of the former intermediaries had been abolished. The gains from this were part psychological, part administrative in that a layer of revenue collectors was eliminated, and part social in that the abolition of the zamindars also abolished certain of their feudal social rights within the village. The immediate economic benefits to the peasant were much smaller; as pointed out, payments by the peasant would not decrease initially since they would now go to the state directly as land revenue. The legislation yielded no direct gains in output, since the problems of raising output are far more deepseated; but it would hopefully improve incentives where it strengthened the rights of cultivators in their land.

These reforms dealt only incidentally with a whole series of other landownership problems. In those states in which intermediaries did not exist intermediary abolition legislation was of course not applicable. Land tenure, size of landholdings, limits on rent, and the condition of agricultural labor were still problems. The Center has recommended legislation to the states in many of these fields—tenant rights, ceilings on land holdings and rents, and consolidation of plots of land. Many of the states have passed legislation in the first two fields, and two of the states have

[2] For example, in Uttar Pradesh State, although only 30,000 zamindars owned 42 per cent of the land and paid annual land revenue in excess of Rs 250, there were almost 2 million who owned 58 per cent of the land and paid less than Rs 250: over 80 per cent paid less than Rs 25. See W. C. Neale, *Economic Change in Rural India,* Part IV (New Haven: Yale University Press, 1962); also F. J. Moore and C. A. Freydig, *Land Tenure Legislation in Uttar Pradesh* (Berkeley and Los Angeles: University of California Press, 1955) for an analysis of the politics of that state's legislation.

passed legislation to expedite consolidation. Nothing has been done with respect to conditions of agricultural labor. In all the areas in which legislation has been passed there has been a time lag in adjudicating legal problems. The effects of the land tenure and the landholding and rent ceiling legislation have been diminished in the short run at least by outright evasion of the laws or by loopholes in them that permit landlords to resume landholdings in cases where they cultivate the land themselves, or let them divide the land among members of their families. Furthermore, the question of the relationship between legislation on size of holding and ceilings and either greater efficiency in use of land or total output of agricultural produce has certainly not been resolved.[3] It is very clear, however, that legislation in this entire area, as well as its implementation, is a major political issue within the states because it vitally affects the property rights and incomes of economic power groups, especially the larger landowners who are normally members of the dominant castes in an area.

Central government advice has also far preceded actual policy with respect to coöperative use of land. There has been a long development of Congress attitudes toward coöperative farming, but in 1959 the Congress party at its Nagpur meeting formally adopted a resolution favoring "coöperative joint farming" as the future agrarian pattern. This was defined roughly to mean joint farming with pooling of land and its centralized management, but the peasants were to be permitted to retain their property rights in the land and receive ownership dividends. The expansion of this coöperative system was to be voluntary, but it was hoped that it would be extended to the entire country within three years. The 1959 resolution largely reflected the influence and policy of Prime Minister Nehru. Regardless of its merits or demerits, the resolution has simply not been

[3] For a recent review of the effects of the legislation see V. M. Dandekar, "A Review of the Land Reform Studies Sponsored by the Planning Commission," *Artha Vijnana,* Vol. IV, No. 4, December, 1962. For an earlier analysis, with a strong point of view in favor of "land to the tillers" see D. Thorner, *The Agrarian Prospect in India.* See also Wolf Ladejinsky, "Agrarian Reform in Asia," *Foreign Affairs,* XLII, No. 3 (April 1964), pp. 445–460.

A recent study of the experience of the intensive agricultural program in Thanjairu District in Madras State, one of the "package programs," made some interesting comments on the relationship between types of landholding and willingness to step up agricultural cultivation. "It was . . . found that improved methods of cultivation were more widely adopted by owner cultivators who form 43 per cent of the Package ryots (peasants). Owner-cum-tenants comprising nearly 25 per cent of the Package participants are also enterprising, but show a tendency to concentrate more on their own holdings than on those taken on lease. Among the lease-deed tenants accounting for another 25 per cent of the Package farmers, those who pay rent on a fixed basis, evince some enthusiasm for intensive cultivation. Tenants paying rent on crop-sharing basis and the oral lessees (seven per cent of farmers) are generally disinclined to exert themselves unduly over stepping up agricultural production." *Hindu Weekly Review,* May 4, 1964, p. 16.

implemented. The states have not taken the steps to carry out the program; and it is noticeable that the Third Five Year Plan was extremely unspecific in the program it set forth with respect to coöperative farming. The enthusiasm of the peasants for coöperative farming is apparently very limited. The approval of the resolution in 1959 was itself the direct occasion of the starting of the Swatantra party, which hoped for support of the peasants with larger landholdings; within the state Congress parties and governments, which have the responsibility for introducing and implementing the legislation in this field, the groups opposing legislation have been sufficiently strong to prevent any action being taken. Given the political position of various peasant groups within Congress this is not surprising.

Although the program on farming coöperatives has not been implemented, various types of service coöperatives have been extended far more successfully. The service coöperatives perform the function of assisting individual peasants in such areas as purchasing, marketing, credit, and even to processing of certain types of crops. To some degree they take the place of existing moneylenders and traders, and these have been threatened by their growth. However, although the new institutions have grown rapidly their influence is still relatively small; furthermore, many of the moneylenders and traders are also peasants and might in this respect gain from expansion of the service coöperatives. In fact, there appear to be close relationships among the growth of the new coöperatives, the local political leadership, and the position of the local dominant subcaste. Members of the dominant caste who also hold major political power within the local and state Congress parties have gained by these institutions. They have gained economically by access to credit, fertilizers, seeds, and implements, they have gained politically by control of a major source of influence and patronage, and they have gained socially by an improvement in their status as a result of their positions in the new institution. They have at times been able to supplant the old moneylenders in function and power. The chief minister of West Bengal stated in 1963:

> Rice and paddy stocks are now in the possession of rich farmers who have emerged as the most powerful group in the rural sector. They have supplanted the traditional moneylenders and, by behaving as enthusiastic cooperators, have been able to corner government loans and assistance.[4]

The spread of these service institutions has thus been encouraged by the shifting of political power in rural India to a broader group than heretofore; as a result the institutions have been favored, have grown rapidly,

---

[4] "Failure of Cooperative Policy Admitted," *Times of India,* August 16, 1963.

and have been used by the dominant castes and rural groups to expand their output and increase their economic power.

AGRICULTURAL TAXATION

A major area of potential policy conflict among the various groups of society is taxation policy. There has been wide consensus on the desirability of a development program, and there has been general agreement on a willingness to raise additional taxes to finance this. From the 1950/1951 to the 1960/1961 budget, the level of total tax revenue rose from Rs 6.2 to 12.2 billion. Of this the great increases have been in internal, indirect taxes; union excise duties, state excise duties, and sales taxes have risen from 27 per cent to 46 per cent of the total tax revenues.

Although there is no record of the changing incidence of indirect taxes over this decade, there is a study of it from 1953/1954 to 1958/1959 on the rural and urban sectors. This indicates a somewhat higher impact of the indirect taxes on the urban sector, but the difference is slight. Table 13 summarizes the results of this study.

TABLE 13
INDIRECT TAXES AS PER CENT OF CONSUMER EXPENDITURES

Group	1953/54	1958/59	Ratio of Percentages 1958/59 Divided by 1953/54
All households	3.6	5.7	160
Rural households	2.9	4.4	150
Urban households	5.9	9.3	158

SOURCE:
GOI Ministry of Finance, *Incidence of Indirect Taxation, 1958–59* (New Delhi: Department of Economic Affairs, 1961); R. N. Bhargava, *Indian Public Finances* (London: George Allen & Unwin, Ltd., 1962).

With respect to direct taxation there is little reason to believe that the proportionate burden borne by agriculture has increased over the decade; it is more likely that it has decreased since, of the main direct taxes, neither the land revenue payments, which include payments formerly made to intermediaries in the early years, nor the agricultural income tax have shown substantial increase over the decade. The main direct tax on agriculture, the land revenue tax, which has not been adjusted since the war, declined from 4.5 per cent of the net value of agricultural output in 1938/1939 to less than 2 per cent of net agricultural output in 1960/1961.

The agricultural price index has risen by about three times during the period 1939–1960 resulting in a decline in the land revenue burden. . . . The average value of production of cereals and cash drops . . . per hectare increased four-fold [from 1934–38 to 1954–58]. . . . The increase in the crop value may be partly attributed to higher yield [of crops].

The money payments due have not been adjusted to reflect the higher prices; likewise the higher yields are not reflected since the states have not made any recent revision of estimated land outputs, which provides the basis of the land revenue system. Although some of the states have agricultural income taxes, these have many defects and their extension has not been advocated. As a result, Ashok Mitra concludes that in the period 1950–1958, while per capita agricultural income averaged about 40 per cent of nonagricultural income, the per capita tax paid by the agricultural population was only about 12 per cent of that paid by the nonagricultural.[5]

There have been many suggestions both to raise the land revenue payments of the peasants and to introduce an element of progression in the system by relating payments to the size of farm, but none has been adopted. In the budget proposals for 1963/1964 the central government proposed a compulsory deposit scheme under which those peasants paying land revenue would be required to deposit with government offices a sum equal to 50 per cent of the land revenue they paid in 1959/1960, or half of the average land revenue of Rs 3 per acre. The proceeds of this compulsory deposit would have gone to the states. However, this indirect proposal to tap a major, potential source of additional revenue was withdrawn after protests from the state governments. This withdrawal is an indication of the power of the peasant groups in both the state and national Congress parties.

The entire discussion may be summarized by looking at the change in the flow of taxes and expenditures between the agricultural and nonagricultural sectors—assuming no shifting of taxes between the two sectors. A comparison of this flow in 1951/1952 and 1960/1961 is summarized in Table 14. Money taxes, unadjusted for price changes, collected from the agricultural sector rose by approximately 80 per cent in 1960/1961 compared with taxes collected in 1951/1952. At the same time annual government money expenditures covered by taxes spent on the agricultural sector rose by 180 per cent between those two years. The net inflow of tax-financed expenditures—sectoral expenditures cov-

[5] On land revenue taxation see A. Mitra, "Tax Burden for Indian Agriculture," *Perspective,* No. 2 (June, 1961), especially pp. 8–13; A. M. Khusro, "Land Revenue: A Plan for Progression," *Times of India,* April 18, 1963. This latter article contains the quotation given.

TABLE 14
INTERSECTIONAL TAX-EXPENDITURE INCIDENCE IN INDIA, 1951/1952 and 1960/1961
(million rupees)

	1951/1952			1960/1961		
	Agricultural Sector	Nonagricultural Sector	Total	Agricultural Sector	Nonagricultural Sector	Total
Tax and other receipts [a] from:	2,391.0	5,513.1	7,904.1	4,334.7	8,904.7	13,239.4
Expenditures in:	3,101.2	3,550.4	6,651.6	8,691.0	9,631.8	18,322.8
Net benefits (expenditures minus receipts)	710.2	–1,962.7	1,252.5 [b]			
Adjusted expenditure						5,083.4 [c]
Expenditure covered by taxes				6,279.8	6,959.6	13,239.4
Net benefits (adjusted expenditure minus receipts)				1,945.1	–1,945.1	0

Notes:
[a] Net contribution from public enterprises.
[b] Surplus.
[c] Deficit.
SOURCE:
Harold M. Groves and Murugappa C. Madhavan, "Agricultural Taxation and India's Third Five-Year Plan," *Land Economics*, XXXVIII, No. 1 (February, 1962), pp. 58–59.

ered by taxes minus sectoral receipts—into the agricultural sector, and thus the rural areas, increased from Rs 710 million in 1951/1952 to Rs 1,945 million in 1960/1961, or by about 175 per cent. At the same time the net tax outflow from the nonagricultural sectors, that is, the more urban ones, remained about constant at Rs 1,950 million in both years. (The government had a surplus in 1951/1952.) If we adjust for price increases of approximately 25 per cent over this period, based on both wholesale price and cost of living indexes, the greater net inflow of real tax resources into the agricultural sector was on the order of 150 per cent.[6]

Very little is known about the flow of private funds and resources between rural and urban sectors.[7] With a tax policy that results in taxes lagging behind expenditures in the agricultural sector, there has been a steady flow of resources through the government from the nonagricultural, to the agricultural sector. Under such circumstances it is not surprising that the shift in the structure of the Indian economy from agriculture to nonagriculture has been lagging behind both hopes and plans.

FARM PRICE POLICY

There is little evidence that the government has adopted a conscious price policy to influence the relation of farm to nonfarm prices in order to shift terms of trade from one to the other. During the decade there had in fact been little change in the terms of trade between these two groups or commodities.

This absence of a policy to shift relative prices may reflect the opposition of two sets of ideas and two political forces. It has been argued that a shift in trade in favor of industrial products would result in a flow of resources to the industrial sector from the rural sector; on the other hand, it has also been advocated that relatively higher prices for farm products will stimulate greater output of farm commodities. These poli-

[6] H. M. Groves and M. C. Madhavan, "Agricultural Taxation and India's Third Five-Year Plan," *Land Economics,* XXXVIII, No. 1 (February, 1962), pp. 57–60. If expenditures of Rs 5,080 million that were not financed by taxes in 1960/1961 are included, the increase in total government expenditures for agriculture between the two years is 180 per cent, and the total net inflow (however financed minus agricultural taxes) into agriculture rises by over 500 per cent. Unfortunately, the authors of this article have not explained their method of sectoral breakdown in any detail.

[7] Since writing this chapter, I have seen the *Economic Survey of Asia and the Far East, 1964* (Bangkok: United Nations, 1965). Intersectoral financing between the agricultural and nonagricultural sectors in India is analyzed in pp. 55–67 (esp. Tables II-22 and II-A-3), and this indicates that there was a much larger net inflow of government and private resources into the agricultural sector in 1957/1958 than in 1951/1952. This of course supports the theme of my chapter.

cies appear directly contradictory: if the first is correct, the second would be incorrect; the first policy might lead to lower output of farm products. Without at this time entering into the merits of these alternative policies, which are discussed in Chapter 12, we may here assert that the first policy would antagonize large and politically potent rural groups; the second would antagonize and raise the cost of living for urban factory workers, trade-unionists, and white-collar workers, who are the most volatile political group in India. Under these circumstances the government has temporized and has adopted neither alternative. In any event price policy has not been used as a means either to transfer resources from agriculture to industry, or to encourage higher farm output.

INDUSTRIAL POLICY

PUBLIC AND PRIVATE SECTORS

The most publicized—at least within the United States—political conflicts within the Congress party arise over the question of government control over industry, and the path toward "the socialistic society." These are questions and issues largely reflecting the intellectuals' influence on policy, their justified worries over the glaring inequalities that characterize Indian society, and their desire for rectification of these inequities. They reflect a bias set in part on the basis of caste factors, but also derived from the somewhat questionable past of private businessmen. The policy was seen as a method for achieving greater equality, curbing the businessmen, and providing a vehicle for the accumulation of capital.

In fact there has been very little nationalization of existing enterprises; the major cases are the internal air lines and the life insurance companies. The government has of course started major industrial works—the three public sector plants in the steel industry, several machine-tool plants and heavy engineering plants, various chemical plants, and coal mines. However, the problems in the public industrial sector are such that its expansion to new industries has been limited. In some states, enterprises have even been sold to private businessmen. It is noticeable that the government has till now successfully resisted the political demand to nationalize banks on the grounds that such a step would have little benefit and would create many problems of management and of business confidence in India and overseas.

Although not much has been done to nationalize existing enterprises, the serious scarcities and bottlenecks, especially of foreign exchange but also of real resources, combined with the distrust by government leaders and members of the bureaucracy of the businessmen, have contributed to an especially complex network of direct administrative controls. This has made it difficult for the private businessmen to function in a socially productive manner, and the government administrators responsible for the controls find it hard to administer them speedily and efficiently. Mostly these controls have reflected the political strength of the urban intellectuals and the bureaucracy, who work in a general environment in which the private businessman is of relatively low repute and status, so that the controls have a broad base of political support and acquiescence by the larger public. There are also by now strong vested interests, both among those businessmen, who gain by the controls and the resulting black markets, and among the administrators carrying out the controls, in continuing them.

It is becoming clearer to governmental and nongovernmental officials and economists that many controls are defeating their broad purpose of development and can be removed; at the same time it is being realized that a rational price system or even uncontrolled prices in more areas than at present, together with proper fiscal policies, may do a better job in many fields of allocating resources than direct physical controls. The influence of private business groups may also be increasing within the party on types of policy issues, such as controls, which can adversely affect businessmen. Furthermore, as private businessmen have been carrying out their industrial plans their reputation has improved somewhat, but it is still low.

Businessmen are becoming more aware than ever before that they are affected by general policies and not just by administrative policy. At the same time their financial importance within the Congress party has increased, and the Swatantra party has arisen as an opposition party committed to reducing controls. This party has some business support and has sought to attract more on this issue. But generally controls are not a major political issue at this time; they can therefore be handled with some degree of pragmatism, and some controls have been relieved. If this issue becomes a major political or ideological issue, the relatively small number of businessmen with their low repute, probably will find it difficult to mobilize support. This pragmatic issue over use of controls is far more significant in India than the issue of conflict between the private and the public sector, which has been stressed in the United States. The latter is not, in my opinion, a real issue. It has not in fact adversely affected private business, which has gained greatly by development; and there is now a mix of private and public ownership and management —government officials or their family members entering business, or sponsoring private businesses—that makes this ideological issue unimportant.

LOCATION OF INDUSTRY

A main area in which the increasing political power of the states makes itself felt is in the location of industry. This is an area of course in which it is difficult to prove policy as a result of state pressure. However, one of the major political problems in India is the tightening of the unity of the country. This problem was more obvious before the Chinese incident of 1962; its clearest manifestations in the South centered upon the DMK agitation in Madras and in the North, upon Kashmir, where the issue is an international one. One of the main weapons that the Center can use to strengthen this unity is an economic policy that ties the various sections of the country more closely together and provides the various states with tangible evidence of its interest. In the case of Kashmir, the relatively heavy development expenditures have been aimed directly at showing national interest in that state. At the same time the state leaders give evidence of their power and importance within the nation by their ability to acquire industrial projects, and this plays an important political role within the states. These considerations unquestionably played a role in the decisions to locate the three new public-sector steel mills in separate states—West Bengal, Orissa, Madhya Pradesh: the fourth proposed mill will be located in Bihar.

One of the major recent arguments arose over the location of a refinery to process Assamese oil. The initial decision was to set up the refinery in Bihar, but subsequently, the decision was changed to set up two refineries, one in Assam and one in Bihar. The change in decision unquestionably reflected strong political pressure from Assam. Similar factors have affected policies with respect to major investment in various ports in Gujerat, Orissa, and West Bengal.[8]

These are some of the more obvious cases of the influence of state political pressures on development decisions. Further examples can be found with little difficulty. In the light of the political situation within India this is not surprising and may well be desirable. India is clearly not peculiar in this regard. For example: decisions with respect to location of defense installations in the United States, where the unity problem was largely solved in 1865, are also strongly influenced by political factors. It is difficult to measure precisely the economic costs and benefits of these decisions, which depend on an uncertain future. However, in India, with its great scarcity of capital resources and the necessity for raising output as rapidly as possible, the effect of grossly uneconomic allocation of re-

---

[8] On the refinery case, see M. Weiner, *The Politics of Scarcity*, pp. 205–207; on Kandla port see D. R. Gadgil, *Planning and Economic Policy in India* (Bombay: Asia Publishing House, 1961) p. xiv.

sources is to sacrifice future economic benefits by raising investment requirements, by lowering rates of return, or by delaying the flow of output. Therefore, although these decisions are understandable and may be desirable on noneconomic grounds, it is preferable that choices be made with conscious recognition of the trade-off between economic gains and political or social gains and with an understanding of the political benefits to be achieved. Instead they are often made on a purely ad hoc basis in response to immediate local political pressures.

Within states, conflicts often arise among various private industrial groups. Many of the states would like to get additional investments by national business groups, of which the Marwari group is probably most important; on the other hand, businessmen from a state urge that local inhabitants should be given priority with respect to more profitable or more desirable investment and job opportunities, and out-of-staters should be discouraged. This is a conflict of interest among industrial groups, and it takes a political form since it requires a decision by the state government; the solution has varied from state to state and even by case. This conflict is exaggerated by the central government policy of sharply restricting entry into many industries and requiring licenses and permits, so that the denial of a permit to any group may make it impossible for it to enter the industry or to expand.

All the above examples point up key economic policy areas that are strongly influenced by the political power of various elements within the Congress. Decisions in these policy areas strongly influence the supply of funds for development, the allocation of investment funds among sectors, and the policies adopted to achieve desired output goals in the economy. The overall effect of the shift in political power has been to strengthen the influence of rural or small-town groups within the Congress. This sector of Indian political society is less interested in industrial development as a goal than urban groups are, and it is also more influential within state governments and parties than urban groups, whose main influence is in the central government. The effect of the shift in power has been to encourage policies that are more conducive to investment in the rural rather than the industrial sector, to a lagging process of resource mobilization from the agricultural sector, and to a wider and uneconomic dispersal of industrial investment among the states. These policies have probably contributed to reducing the rate of overall Indian economic growth and to slowing the process of sectoral shift within the economy.

8

Gains and Losses of Development—Rural India

INTRODUCTION

The following chapters present a statistical analysis of the gains and losses from economic development in the past decade. These conclusions are again based on scraps of statistics supported by impressions. Indian statistics on income distribution over time or in any one year are either absent or extremely meager. The recent report of the Committee on Distribution of Income and Levels of Living, the Mahalanobis Committee, concluded that "the required data are not available at present for a direct study of the question of income distribution . . . and no firm conclusions on the subject can be drawn."[1]

The analysis of the overall economic changes during the decade 1951–1961 indicates a general improvement in the well-being of the population, with the rate of increase in total output keeping ahead of the rate of population growth. From the theoretical analysis it would be expected that there would be some relationship between the shifts in political power over the period and the distribution of economic gains.

A major political result of democracy was the widening of the base of political power in India so that numbers became a source of votes and influence. This was accompanied by some widening of economic power which can be associated with a widespread distribution of the gains of economic development. I stress association rather than cause, since, as I pointed out earlier, factors other than the political factors will influence the observable changes in income distribution.

In the rural sector the land reform legislation, admittedly not up to its more idealistic expectations in its effects, did distribute land more widely and reduced the position of the largest landholders. It is probable that the peasants with medium-sized landholdings have gained the most by the rural changes. In many areas these have been the peasants with a tradition of working their own land. They have been the ones with sufficient economic resources to take advantage of the introduction of new

[1] *Mahalanobis Committee,* p. 10. See also pp. 10–12 for a discussion of the weaknesses of existing data on income distribution.

techniques and inputs, and they have been the ones with the greatest political power at the state and local level and therefore dominant in new local institutions. The peasants with smaller landholdings have gained somewhat economically and socially, both through the spread of educational institutions and by the increase in the number of new nonagricultural jobs in government and industry. The agricultural laborers have gained the least, for they had little political or economic power—in many areas even little incentive—to take advantage of the changes. Although many members of the untouchable castes are concentrated in this lowest sector, there are many outside of it; and in relative terms, the new educational opportunities and the privileged position for members of these castes in government have contributed to their improvement in status. Apart from the income data which tend to support this thesis, the census data on urbanization in the 1950's also indicate some improvement in rural conditions. Contrary to what was expected, the last decade saw a decline in the rate of urbanization compared with earlier decades. There is reason to believe that on the positive side this has been associated both with the improvement in rural incomes and the increased welfare expenditures in the rural sector; on the negative side, the increased open urban unemployment and the more glaring discomforts of ever more crowded urban areas have reduced some of the city's attractions.

In the urban sector, there has been a general increase in employment opportunities for all groups both because of the expanding scope of government and as a result of economic development. Those middle-class and industrial groups that have the knowledge, the resources, and the political contacts to take advantage of the development programs in industry and commerce have gained the most; this includes entrepreneurs, professional people, and management groups. The factory workers also gained significantly, at least until 1961. The increased demand for their skills and the sensitivity of the Congress party to the union demands have contributed to this.

The lower middle-class groups in commerce and all white-collar groups in government have gained the least. The numbers in these categories have increased so that they have had a total income gain, but the per capita incomes of the members of these groups have tended to lag—a lag that is both relative to other groups and, in some cases, even absolutely in relation to the cost of living. This relative economic loss has been associated with the fact that many of these white-collar workers were Brahmins, and the loss of political power and status by Brahmin groups has carried over into their economic influence. The lower white-collar, clerical sector has also had probably the greatest oversupply of job applicants, and this has obviously contributed to the economic decline of its members. To the extent that Brahmins have moved to managerial and entrepreneurial positions, they have shared the gains of the middle-class groups in those positions.

A very serious and disquieting element in this picture of general gain is the rise in both total and educated unemployment. The total number of registered unemployed rose from 330,000 in 1951 to 700,000 in 1955 to 1.8 million in 1961. The registered figure is a very large underestimate since only about 25 per cent of the total unemployed are believed to register, but it probably does measure open urban unemployment best. Within the total figure the number of registered educated unemployed (those with the equivalent of at least a high school education) rose from 216,000 in 1955 to 600,000 in 1961; again only about 40 per cent of the educated unemployed are believed to register.[2] This increase of open unemployment, especially among the educated, is a sign of the tenuousness of the past economic achievement as well as a potentially significant political threat.

Three recent efforts have been made to examine changes in the distribution of expenditures and income during this period. Since the results support my general analysis, they may be used as a summary of national changes before I proceed to the detailed analysis of sectoral changes.

A preliminary survey of the change in the distribution of expenditure among expenditure classes over the period 1951–1957 was based upon National Sample Surveys during that period.[3] The main finding of this analysis is that the distribution of expenditures showed little change over this period. The share of the 10 per cent of the rural households in the lowest expenditure classes as a percentage of total expenditures rose very slightly, whereas that of the similar 10 per cent of the urban households remained about constant. The high margin of statistical error with respect to the top 10 per cent of the households in both the rural and urban sectors makes the results for this group relatively doubtful; but for what they are worth, the share of the 10 per cent of the urban households in the highest expenditure classes as a proportion of total expenditures rose somewhat, whereas the share of the similar 10 per cent in the rural household population remained roughly constant. Within this upper expenditure class, the number of people in the top 5 per cent class rose from 0.9 to 1.0 per cent of India's total household population, whereas the share of the same group in total national expenditures rose from 3 per cent to 5 per cent. The conclusions from this analysis are: first, that there may have been some slight improvement in their shares of total expenditures by both the lowest 10 per cent of India's population and the

[2] See Appendixes C and D for these figures.

[3] Indian Statistical Institute, "Preliminary Study of April 7, 1961" (mimeo).

top 5 per cent; and second, that these shifts, if they did occur, were very minor within the total structure of expenditures.

The Reserve Bank of India made a study of changes in income distribution during the short intra-Plan period from 1953/1954 to 1956/1957.[4] In this period prices remained approximately stable, so there was no serious need for price adjustments to make a real comparison. The method used was to apply National Sample Survey data with respect to expenditure distribution to the national income data; and such additional income data as were available were also fed into the analysis. In overall terms the Reserve Bank concluded that the changes in income distribution during those four years were slight. In both years 95 per cent of all Indian households received incomes of less than Rs 3,000 per year, and this overwhelming proportion of the population is estimated to have received 81 per cent of the national personal income in 1953/1954, and 80 per cent in 1956/1957. At the highest income level, those households receiving incomes of Rs 25,000 per year made up only 0.2 per cent of the total number of households in both years; and received about 5 per cent of the total personal income. These proportions are all about the same if disposable income, that is, after direct taxes, is used rather than personal income.

Within this general pattern the statistical data showed that the proportion of income received by rural households in the income group below Rs 3,000 per year—including the peasants with medium and small landholdings—rose slightly, while that of urban households in the same income group fell somewhat. In the income group above Rs 3,000 per year, urban households gained relatively and rural housholds lost slightly.[5] This improvement in the relative position of the urban high-income group during these four years, it is considered, explains part of the supposed rise in the proportion of savings to total disposable income of the entire household sector from 5.3 per cent to 7.5 per cent during the period.

The Mahalanobis Committee, on the basis of both these studies and certain special studies, concluded that "available estimates and data suggest no significant change in the overall distribution of incomes, though they do indicate a slight probable increase in inequality in the urban

[4] "Distribution of Income in the Indian Economy: 1953–54 to 1956–57," *Reserve Bank of India Bulletin,* XVI, No. 9 (September, 1962), pp. 1348–1363.

[5] This apparent loss in the relative position of the rural group above Rs 3,000 income would probably reflect the decline in the position of the largest farmers as a result of the land reforms, especially the abolition of intermediaries. In the short period from 1953/1954 to 1956/1957 this group would not have been able to recoup their losses from investments elsewhere. It should be realized that the data with respect to rural incomes are especially weak, and the sample for high income farmers is very small, so the result may reflect the statistics rather than the events.

sector and some reduction in inequality in the rural sector. . . . [In] view of the inadequacies of the data used for comparison purposes it is not possible to be definite about this conclusion."[6]

These results, for what they are worth, tend roughly to support the previous analysis of the gains from development; they also point up again most clearly the minor effects of the past development effort upon the structure of the economy. With little change during the decade in either the structure of occupations or in the rural and urban distribution of the population, it is not surprising that the structure of income distribution also showed relatively little change. It moved neither very much toward greater inequality, which might have contributed toward greater voluntary personal savings, nor toward the goal of increased equality set forth in the Constitution. At the same time, the stability of the economic structure of the country was accompanied by a great degree of political stability within India as the Congress party maintained an overwhelming political leadership during the entire period. It seems that the actual economic policy of the decade—economic development without major structural change—contributed to that stability. But it may have high risks for future development, especially if the aspirations for improved economic well-being have in fact become more general. In the next section of this chapter and in Chapter 9 and Appendixes C and D, I present the statistical evidence, by rural and urban sectors and by class groups, for the generalizations in this section. Chapter 10 will discuss caste and communal economic gain and losses since Independence.

RURAL INDIA

Chapter 6 presented a brief overview of the economic achievements and structural changes in the Indian economy from 1951 to 1961. This is a necessary preliminary to estimating the distribution of the achieved economic gains among those various groups within Indian society that were identified in Part II.

The major division in Indian society that has not changed appreciably within the decade is that between rural and urban sectors; this division obviously reflects the differences between agricultural and nonagricultural activities. Within the rural sector I earlier identified caste differences and economic class differences in terms of size of landholdings and commercialization of farm activities. Within the urban sector major stress was laid upon economic class differences, and various urban middle-class and working-class groups were identified, with recog-

[6] *Mahalanobis Committee,* p. 23.

nition of the influence of caste factors within urban areas. On the basis of the general income and price statistics available and on the basis of isolated data on numbers and incomes of particular groups from government documents and reports or private research, supplemented by nonstatistical knowledge, it is possible to make crude estimates with respect to the trend of real income changes in and among these groups. However, it may not be possible to aggregate these figures and trends into an income figure for the group. I feel that statistical estimates of the direction of change in the income position of the identified classes can be made with a fair degree of general accuracy. Unfortunately I cannot say the same for the economic position of the caste and communal groups. To a certain degree some of the broad changes in communal incomes can be shown from state income data now that the states represent linguistic groups. With respect to the distribution of gains and losses among caste groups or smaller communal groups within these linguistic areas, it will be almost impossible to make any precise statistical estimates; it may, however, be possible to indicate certain broad trends on the basis of personal knowledge and the research of others. Such changes are very closely related to changes in political power.

DISTRIBUTION OF GAINS BETWEEN THE RURAL AND URBAN SECTORS

From the earlier discussion of the relation between shifts in political power and the gains from development, one would expect at least an absolute improvement, possibly a relative one, in the economic position of the rural population with the shift in power to rural elements since Independence. Table 15 presents from the 1961 Census the absolute figures of urban and rural workers and population—by the same occupational classification as in Table 11—and shows a 38 per cent increase in rural employment compared with only a 15 per cent increase in urban employment. At the same time the total rural population rose by 20 per cent from its unadjusted 1951 base, while urban population rose by 27 per cent from the same base.[7] The different movements of rural and urban populations and job figures are difficult to explain.[8] The major absolute

[7] The "adjusted" urban population was 58 million and the rural population was 295 million in 1951. This base yields a 36 per cent rise in the urban population. Unfortunately no other figures have been changed for the adjustments, so that all the analysis that follows is in terms of "unadjusted" figures.

[8] Some of this rural rise may be definitional and reflect the rise in population rather than jobs. In urban areas I would expect a narrower definition of employment—excluding children—and the figures would thus reflect the rise in jobs. In rural areas most of the greater population living on the land and physically able to work—and in rural areas this might start at ages of 5 and 6—would be considered employed, since they do work. It is noticeable that in the urban areas the ratio of nonworkers to workers is about 2:1, and has in fact risen over the

TABLE 15
BREAKDOWN OF RURAL AND URBAN POPULATIONS BY OCCUPATIONAL GROUPS, 1951 and 1961
(millions)

Sector	Rural Population [a]				Urban Population [a]			
	1951	Per Cent	1961	Per Cent	1951	Per Cent	1961	Per Cent
1. Cultivator	68.8	58	98.0	60	1.8	8	1.7	7
2. Agricultural laborer	26.7	23	30.6	19	1.1	5	0.9	4
3. Mining, forestry, etc.	3.8	3	4.5	3	0.5	2	0.7	3
4. Household industry [b]	—	—	10.0	6	—	—	2.1	8
5. Manufacturing	6.7	6	2.4	1	5.8	25	5.5	21
6. Construction	0.8	1	1.1	1	0.7	3	1.0	4
7. Trade and commerce	3.1	3	3.3	2	4.2	18	4.3	16
8. Transportation, etc.	0.7	1	0.9	1	1.5	6	2.1	8
9. Other services	7.4	6	11.5	7	7.4	32	8.1	31
Total [c]	118.1	100	162.4	100	23.1	100	26.4	100
10. Non-workers	180.7		197.9		39.2		52.4	
Total population [d]	298.8		360.4		62.3		78.8	

Notes:
[a] 1951 figures are "unadjusted" for definitional changes. Between the 1951 and 1961 censuses, 812 swollen villages with no other urban characteristics except population, and totalling about 4.0 million people, were excluded; the 1961 census figure included 483 new places with urban characteristics.
[b] In 1951 occupational figures for workers in "household industry" and "manufacturing" were combined under "manufacturing"; in 1961 they were separated.
[c] Differences in totals due to rounding.
[d] Slight differences in totals from previous tables reflect either slight coverage differences or results of working with percentages and rounding.
SOURCE:
Registrar General of India, "Final Population Totals," *Census of India, 1961,* Paper No. 1 (1962).

and relative employment increases in the rural sector were in the number of cultivators, which may reflect the land reforms. In industry, the increases in household industry especially bulked large; and increases in the services reflected such efforts as community development.

Unfortunately, income data broken down by rural and urban sectors are not available for 1951 and 1961. The only income and employment figures are by occupational, rather than locational, sectors. These are presented in Table 16; but I will later try to translate these into urban and rural estimates.

Table 16 shows a surprising fall in the real output (or income) per worker in the mining, manufacturing, and small enterprise sector. This may be due to defects in employment and output statistics. On the employment side, the failure in 1951 to distinguish small enterprise workers from others combined with the relatively large role of small enterprise, would lead to error. On the output side, P. N. Dhar and S. S. Sivasubramanian have estimated a 32 per cent increase in net output of small enterprises at current prices from 1955/1956 to 1959/1960, in contrast to the official estimate of only a 24 per cent increase.[9] If this income correction is made, the output per worker in this sector would have remained roughly constant. What is also very striking, granted the weakness of the statistics, is the great increase in the output per worker in the tertiary sector (Sectors 3 and 4 in Table 16), which exceeded 20 per cent in the decade in contrast to the relative stability in the other sectors.

The translation of this occupational sector data into a rural and urban sector breakdown is far more difficult, since reliable income information is not available on such a basis. However, a first attempt at such a breakdown was made for the 1952/1953 national income data.[10] I apply the 1952/1953 proportions to the 1950/1951 real national income data on the assumption that changes between these two early years, both roughly at the start of the First Plan, were negligible. In the light of the population movement to urban areas in the decade from 1951 to 1961, and the relatively more important role of employment in "manufacturing" and "other services" in the urban areas, I assume a 2 per cent shift in the

decade; in rural areas the ratio is much closer to 1:1 and has in fact declined. The relative shift of workers from "agricultural laborers" to "cultivators" may also contribute to this, if more members of the cultivators' families work on their own farms. It may of course also reflect improved reporting in the rural areas from 1951 to 1961.

[9] P. N. Dhar and S. Sivasubramanian, "Small Enterprises: Their Contribution to National Income," *The Economic Weekly*, XIV, Nos. 28–30 (Special Number; July, 1962), p. 1167.

[10] S. K. Chakravarti, U. Datta, and V. Srinivasan, "Share of Urban and Rural Sectors in the Domestic Project in India in 1952/1953," in *Papers on National Income*, eds. V. K. R. V. Rao, *et al.* (Bombay: Asia Publishing House, 1960).

TABLE 16
SECTORAL DISTRIBUTION OF EMPLOYMENT AND NATIONAL DOMESTIC OUTPUT, 1951 and 1961
(in 1948–1949 prices)

Sector	1951			1961		
	Number of Workers (millions)	Net Domestic Output, 1950/1951 (billions of rupees)	Output Per Worker (rupees)	Number of Workers (millions)	Net Domestic Output, 1960/1961 (billions of rupees)	Output Per Worker (rupees)
1. Agriculture, animal husbandry, etc.	100.6 [a]	43.4	431	135.1 [a]	59.1	437
2. Mining, manufacturing, small enterprises	14.9 [b]	14.8	994	23.2 [b]	21.1	909
	(13.4) [c]	(14.8)	(1,104)	(21.1)	(21.1)	(1,000)
3. Commerce, transportation, communications	9.4	16.4	1,766	10.6	24.6	2,321
4. Other services	14.6	13.9	952	19.6	23.2	1,184
Total	139.5	88.7	636	188.5	128.0	679

Notes:
[a] Excludes workers in mining, estimated at 0.6 per cent of the total working force in 1951; using this proportion for both 1951 and 1961, the absolute figures for numbers of miners are 0.8 million in 1951 and 1.1 million in 1961. These are included in the second sector.
[b] Income includes income from "mining" and is assumed to include income from "construction." Employment in this sector includes construction workers. See also footnote c.
[c] The figures in parentheses are exclusive of construction workers, or are computed excluding construction workers from the denominator to estimate output per worker.

SOURCES:
Table 11, p. 138; GOI Central Statistical Organization, Department of Statistics, Cabinet Secretariat, *Estimates of National Income, 1948-49 to 1961-62* (New Delhi: January, 1963), p. 22.

origin of income in those two occupational sectors from rural to urban areas during the decade. In other occupational sectors I assume that the rural to urban income proportions remain constant. On the basis of these assumptions, Table 17 presents a breakdown of national income both by occupational sectors and by rural and urban origins for 1950/1951 and 1960/1961.

The income gains in the urban area were relatively greater; the differences in per capita outputs become slightly more favorable—in per-worker outputs, much more favorable—to the urban areas over the ten years. The output figures are estimated at factor cost and thus are pretax figures. In Chapter 7, I indicated an increase in the annual inflow of expenditures after taxes into the rural sector of 1,200 million rupees, while the net outflow from the urban sector remained roughly constant. Thus the post-tax effects of government outlays have been to improve the relative position of the agricultural and rural sectors in comparison with nonagricultural and nonurban sectors, and possibly to eliminate the apparent income shift before taxes in favor of the urban sector.

It would be expected that this absolute improvement in the position of the rural sector has its consequences on the movement of population from the rural to urban sectors. In fact, since the relative importance of employment, investment, and income in the various sectors—especially agricultural and nonagricultural—have shown only minor changes, it can be hypothesized that the rate of urbanization, which normally is associated positively with the speed of industrialization, should also show only minor changes.

With the land reforms and the great increase in the number of agricultural cultivators, with more peasants owning their own plots than before, it can be expected that the desire to stay on the land to cultivate those plots is increased. At the same time, the absolute improvement in per capita output in the rural areas makes it more tolerable for the peasants to remain on the land. Thus the "push" elements from the rural areas for migration have apparently diminished. With respect to the "pull" elements, the statistics show some slight improvement in relative per capita incomes for the urban areas, but the difference over time appears slight. One of the most striking facts was the lag in urban employment behind urban population growth, so that in percentage terms urban jobs grew less rapidly than urban population; and open urban unemployment increased in absolute figures.[11]

Unquestionably, during the decade there was a physical deterioration in the urban areas; overcrowding in housing, education, and health facili-

[11] See Chapter 10 and Appendixes C and D for the actual estimates.

TABLE 17
BREAKDOWN OF REAL DOMESTIC NATIONAL OUTPUT
BY RURAL TO URBAN PROPORTIONS, 1950/1951–1960/1961

Occupational Sector	1952/1953 Proportions (per cent)		1950/1951 Breakdown (billions of 1948/1949 rupees)		1960/1961 Proportions (per cent)		1960/1961 Breakdown (billions of 1948/1949 rupees)	
	Rural	Urban	Rural	Urban	Rural	Urban	Rural	Urban
Agriculture	96	4	41.7	1.7	96	4	56.7	2.4
Manufacturing	47	53	7.0	7.8	45	55	9.5	11.6
Commerce	44	56	7.3	9.3	44	56	10.8	13.8
Other services	37	63	5.1	8.8	35	65	8.1	15.1
Totals			61.1	27.6			85.1	42.9
Total population (millions)			299	62 [a]			360	79
Number of workers (millions)			118	23			162	26
Output per capita (rupees)			204	445			236	543
Output per worker (rupees)			518	1,200			525	1,650
Rural output per capita / Urban output per capita =			46				44	
Rural output per worker / Urban output per worker =			43				32	
Per cent change in output per capita							+16	+22
Per cent change in output per worker							+ 1	+38

Note:
[a] Unadjusted for definitional changes, since neither sectoral employment nor 1952/1953 sectoral income data have as yet been adjusted for subsequent definition changes. This table does indicate an improvement in real output per capita and per worker in both the rural and urban areas.

SOURCES:
S. K. Chakravarti, U. Datta, and V. Srinivasan, "Share of Urban and Rural Sectors in the Domestic Project in India in 1952–53," in V.K.R.V. Rao *et al.* (eds.), *Papers on National Income and Allied Topics* (Bombay: Asia Publishing House, 1960), Vol. I, p. 158; Tables 15 and 16 above.

ties became serious, while these same facilities became more numerous in the rural areas as a result of government programs. This in turn somewhat reflected the shift in political power from urban to rural groups. The effect of these employment and amenities factors would be to diminish the force of the "pull" element of income difference which draws peasants from rural to urban areas. Although there was the previously noted absolute increase in the urban population over the decade—a rise of 36 per cent over the adjusted 1951 base, and 27 per cent over the unadjusted 1951 base compared with a 22 per cent increase in total Indian population —the relative shift in the proportion of rural to urban population was only about 2 per cent at most.[12] This was a much slower rate of increase of urban population and slower shift than had been anticipated before the Census; and it was a slower rate of change than in previous decades since 1921. The absolute increase in the urban population was about equal to that in the 1941–1951 decade, which included a war, Independence, and Partition. This slowing down of the rate of urbanization supports the hypothesis that the shift in political power to rural groups since Independence had economic consequences in terms of the relative investment and employment opportunities in the rural and urban sectors, and thus the lack of sectoral change within the Indian economy. This in turn contributed to a reduced rate of urban population growth. It is probable that this reduced rate of urban growth, even if it was associated with a failure of planned economic hopes to be fulfilled, contributed to the social and political stability in India during the decade. But again the question arises as to the long-run relationship and the tradeoffs between a reduced rate of economic growth and urban shift, and the political stability in India.

DISTRIBUTION OF GAINS AMONG RURAL CLASSES

Agricultural Classes. I will repeat the very rough identification of the three economic groups within Indian agriculture that was made in Part II.

[12] Within this slight relative shift in the total urban population, the largest cities showed a very marked relative increase. All cities of 100,000 or more grew by 48 per cent; and those of 1,000,000 or more by 80 per cent. The proportion of the urban population in cities of 1,000,000 rose from 13 per cent to 18 per cent of the total in the decade; and the proportion in all cities of 100,000 or more from 38 per cent to 44 per cent of the total. Since cities of 1,000,000 or more tend to have a larger proportion of their labor force in tertiary employment than other areas, this growth would reflect the increase in this sector, especially in the government. The potential demands on the economy in the way of social overhead to meet the increasing problems of these great cities are great, and it is a serious question to what extent they can be met economically or will be dealt with politically, given the shift of political strength between urban and rural areas. *Census of India, 1961,* pp. xxxi–xxxvi.

(1) Those peasants with holdings of 15 acres or more. Although these were only a small proportion of the total number of peasants in 1950/1951, the total size of their holdings was a larger proportion of the total cultivated acreage in India than the size of their population to the total. The proportion of the agricultural output marketed by them was relatively high, indicating a greater commercial orientation.

(2) The smaller landholding peasants, with 15 acres set as the dividing line between them and the larger peasants. This line is relatively arbitrary. It may in fact vary from a rice to a wheat producing area and from state to state. This group includes the greatest number of peasants holding the largest amount of acreage, but these peasants are much more subsistence-minded than the first group, and their crop sales are in large part due to economic pressures.

(3) The agricultural laborers, who derive most of their income from work for others although they may have small landholdings. (It is obvious that an income classification rather than a landholding one would be preferable, if data would permit it.)

I assume that the proportionate contribution of each of the above landholding classes to real domestic agricultural output measured in net value-added terms in 1950/1951 was the same as its proportionate contribution to the gross output of farm products in that year. On this assumption Table 18 presents a breakdown of contribution to national income in 1950/1951 by size of holding.

TABLE 18
CONTRIBUTION TO REAL FARM INCOME BY SIZE OF HOLDING, 1950/1951
(In 1948/1949 prices)

Size of Holding Class	Proportionate Contribution to Gross Farm Output	Income Produced in 1950/1951 (Rs billions)
0–15 acres	62.1	27.0
15+ acres	37.9	16.4
Total	100.0	43.4

SOURCES:
Table 3 and Table 16.

The real output produced from agriculture rose by 36 per cent from 1951 to 1961. Unfortunately, "there is very little data in regard to the classification of holdings according to size, their essential characteristics such as tenure and tenancy and the resources available to each major type of farm."[13] And there is little data with respect to output changes

[13] Quoting a speech by Professor M. S. Thacker, Member of the Planning Commission, in *Hindu Weekly Review*, July 29, 1963.

among the types of farms. Some evidence, however, supports the thesis that the larger farmers were the largest gainers from the various agricultural development programs of the decade. This conclusion is based on various village studies and official reports that have been issued during the period. S. C. Dube, after a survey of a community development project in two villages in Uttar Pradesh, concluded:

> Nearly 70 per cent of the benefits [from the agricultural extension program in these villages] went to the elite group [of villagers] and to the more affluent and influential agriculturists. The gains to poorer agriculturists were considerably smaller. Being suspicious of government officials they did not seek help from the Project as often. As this group had little influence on the village and outside, and was in no position to offer any material help in the furtherance of Project objectives, the officials largely ignored it.[14]

Similarly, a review of the experience of several other villages in Uttar Pradesh state found that in the three-year period from 1956 to 1959, the net capital investment per hectare of cultivated land was appreciably higher in a village that was in a community development block than in one that was not. Within either block or nonblock villages the amount of net capital investment was directly related to differences in income and size of holdings, and finally it was found that state aid generally flowed to the relatively better-off cultivators.[15]

Writing of a South Indian village, Beals found:

> Those who make adequate profits and have acquired surplus capital are the large landholders in the Gopalpur region. . . . The kinds of agricultural improvements introduced by those who farm large acreages are those which produce a reasonable yield with the least possible effort. . . . [But] the small farm becomes an economically feasible operation only when the greatest possible yield is produced even though this demands the most possible effort.

And he stressed a tendency toward the introduction by the larger farmers of relatively low-yield crops, which require large acreages and little labor, simultaneously with labor-saving investments and technological changes.[16]

The Rural Credit Follow-Up Survey found that coöperative credit

---

[14] S. C. Dube, *India's Changing Villages* (Ithaca, N. Y.: Cornell University Press, 1958), pp. 82–83.

[15] "Capital Formation in Agriculture—A Case Study in Uttar Pradesh, India," *U. N. Economic Bulletin for Asia and the Far East,* XII, No. 2 (September, 1961), pp. 29–44.

[16] A. R. Beals, *Gopalpur* (New York: Holt, Rinehart and Winston, 1963), pp. 80–81. Beals is not an economist but he has noted a shift among larger farmers from maximizing outputs to maximizing profits; while smaller subsistence farmers concentrate on the former.

facilities were availed of to a far greater extent by the more credit-worthy larger farmers than by the smaller ones. Also, a much larger proportion of the borrowings of the larger farmers were used for capital investments, whereas much of the debt of smaller farmers was for consumption purposes of one type or another. I have already noted the comments that the new credit institutions made it possible for the larger farmers to withhold their output from market for longer periods than previously in expectation of higher prices, and in fact in many places they engage in trading at the expense of traditional traders.[17]

Finally, on an overall basis, the Government of India Study Team on Community Projects and the National Extension Service reported:

> There is a direct relationship between the size of landholding for a group and the proportion of respondents from that group that derive benefit from the particular [programme]. Thus . . . 66 per cent of the larger owner-cultivators, 46 per cent of the medium owner cultivators, and 22 per cent of the small owner cultivators derive benefit from the programme of improved seed supply. The same is found to be true about manures and fertilizers, improved methods of cultivation and pesticides. . . . [The] tenant cultivator figures the least in proportion to his size among the beneficiaries. . . . [If he] has no land of his own he is probably reluctant to invest additionally on . . . facilities unless the landowner shares the cost with him. . . . [It may also] be that he has no land against which he can borrow credit either in cash or kind. . . . In other words, the better off a cultivator is, the more likely is it that he will figure in this list of beneficiaries from project programmes.[18]

A very important additional factor is that the larger farmer is probably better able to take the risks involved in experimenting with new inputs, new outputs, or new technology. Since he has land above that necessary for subsistence, he can take more of a chance of using some of that surplus for experiment; since he has either surplus capital or is credit-worthy for loans, he can risk a loss on an investment. The smaller

---

[17] Reserve Bank of India, *Rural Credit Follow-Up Survey, 1958–1959* (Bombay: Examiner Press, 1961), chaps. ii–vi and xii. For additional supporting references see *United Nations Report of a Community Development Evaluation Mission in India*, August 17, 1959; and in the irrigation field see D. Thorner's article on the Sarda Canal in *Land and Labour in India*.

[18] Quoted by K. N. Raj, "Some Features of the Economic Growth of the Past Decade in India," p. 262; also for an excellent summary analysis see his article, "Changing Outlook in Indian Agriculture," *The Listener*, August 18, 1960, pp. 245–247. See also the supporting article by S. C. Gupta, "New Trends of Growth," *Seminar*, XXXVIII (October, 1962). On the risk factor see H. Singh, "Uncertainties and the Adoption of New Practices in Agriculture," *The Economic Weekly*, XVI, No. 22 (May 30, 1964), pp. 925–927. See also "U.S. Team's Survey of Land Reform," *Hindu Weekly Review*, August 31, 1964, p. 10. This is apparently the Ladejinsky Report described in *The Economic Weekly*, August 29, 1964, p. 1428.

peasant, producing for subsistence and with no surplus of resources, is much less able to take such risks and earn the gains possible from taking them.

On the basis of this admittedly very scattered evidence, it is likely that the larger peasants were better able to take advantage of the new facilities to increase their output and generally gained more than the smaller ones. However, if a comparison over time is to be made, the question arises as to what extent the size of holdings and thus the proportions within each size group have changed over the past decade. On this question data are lacking, and one must rely on what are largely intuitive judgments. The land reforms, especially those imposing ceilings, have probably reduced the sizes of the above-ceiling holdings—those above 30 acres in most states. The rise of almost 50 per cent in the number of agricultural cultivators, shown in the 1961 Census, supports the belief that the number of small landholders increased; and part, at least, of this disproportionate increase was probably a result of the land reforms, especially the ceilings. This is in spite of widespread evasion of the ceilings by distributing lands among family members. Thus the land redistribution effects of the reforms, purely in terms of numbers, would have operated to increase the number of smaller farmers. The Mahalanobis Committee, using National Sample Survey data, found a slight trend toward redistribution of agricultural landholdings between 1953/1954 and 1959/1960. In the first period the top 10 per cent of rural households owned 58 per cent of all lands belonging to households; in 1959/1960 this group owned 56 per cent of the lands. The bottom 20 per cent of rural households owned no land in both periods.[19] Combined with this effect, the abolition of the intermediaries would have raised the pretax income of the smaller peasants relative to those of the larger intermediaries, and this too would be reflected in the proportionate distribution of income gains between larger and smaller peasants over the decade.

I arbitrarily assume that the peasants with holdings below 15 acres increased the total output from those holdings by 25 per cent because of new techniques, additional inputs including labor, and the incentive of direct ownership. This is somewhat more than the 20 per cent increase in rural population so that their per capita output rose. I also roughly guess that they received another 5 per cent increase in their total income as a result of the distributional effects of the land reforms noted above. Thus the total increase in value output for the group of peasants owning 0–15 acres would be 30 per cent, or Rs 8.1 billion over the 1950/1951 figure as expressed in constant prices. On the basis of this estimate for

[19] *Mahalanobis Committee,* p. 29.

these peasants, the increases in output for the larger peasants would have been Rs 7.6 billion over the 1950/1951 figure. This is summarized in Table 19.

TABLE 19
ESTIMATED CONTRIBUTION TO RURAL FARM INCOME BY PEASANTS ACCORDING TO SIZE OF HOLDINGS, 1960/1961
(billions of rupees)

Holding Group	Real Income Produced (1948/1949 Prices) 1950/1951	1960/1961	Percentage Change Over Decade
0–15 acres	27.0	35.1	+30
15+ acres	16.4	24.0	+46
Total	43.4	59.1	+36

SOURCE:
Table 18.

Before it is possible to estimate the gain of either group of cultivators, it is necessary to subtract from the above figures the wage payments to farm laborers. How did the agricultural laborer share in these gains in agricultural incomes? Although some statistical data are available they are quite undependable, and any estimate will have to be impressionistic. The two Agricultural Labor Enquiries conducted for 1950/1951 and 1956/1957 found a decline of about 10 per cent in the incomes of agricultural laborers in the seven year interval around which they were conducted. However, after examination of the enquiries it was concluded that as a result of changes in the definition of laborers, changes in the price deflators used for estimating payments in kind, and other statistical changes, the one thing the two enquiries should not be used for is to compare the income position of the laborers in the two years. Furthermore, the enquiries did not limit themselves to landless laborers, but included laborers who had land and earned income from their land.[20]

The enquiries did find a decrease in the number of agricultural laborers between 1950/1951 and 1956/1957. This could well indicate an improvement in their position since those laborers with land may have benefited both from the general improvement in agriculture during this period and by acquiring more land as a result of the reforms, while remaining small landholders. The effects of both of these changes would have been to reduce the dependence of such peasants on the income they

[20] On this see GOI, *Report on the Second Agricultural Labour Enquiry, 1956–1957* (Simla: 1960); V.K.R.V. Rao (ed.), *Agricultural Labour in India* (Bombay: Asia Publishing House, 1962), especially the editor's Introduction and the contribution by K. N. Raj; D. and A. Thorner, "The Agricultural Labour Enquiry," *Land and Labour in India.*

earned from agricultural labor, and thus they would have left the ranks of agricultural laborers as defined by the enquiries. Although the 1961 Census estimated an absolute increase in the number of agricultural laborers from 1951, it also showed that the relative number of agricultural laborers declined from 20 per cent to 17 per cent of the total number of agricultural cultivators plus laborers. Although the rural population as a whole increased by about 20 per cent, and the agricultural cultivators by 43 per cent, the number of agricultural laborers rose by only 14 per cent, indicating a relative outflow from that occupation.[21]

From the scattered data available with respect to daily money wages of agricultural laborers during this decade there is little evidence of any rising trend. Their wages fluctuate so greatly from year to year, from season to season, and from locality to locality depending on immediate and local demand and supply conditions, that it is difficult to discern any long-term pattern from the raw data; and the data are impossible to adjust without far more information than is available on local prices, types of payment, and so on. In part this apparent constancy of money wages probably reflects the absolute increase in the number of such laborers from 1951 to 1961 when there probably already was substantial surplus labor in agriculture. The increase in the absolute numbers in turn may have been a result of both the growth of the rural population and the land reforms. As previously noted, there was an increase in the number of evictions of tenants under the land reform laws as landlords resumed land for personal cultivation, and family operation of farms increased in order to prove personal cultivation. A further complication arises from the fact that although money wages remained constant, payments in kind, which are jajmani payments and thus a proportion of output, may have increased as total output rose.

I shall therefore assume, from these contradictory tendencies, that the real income per agricultural laborer did not deteriorate over the decade, but also that it did not improve. There may have been nonincome gains or losses that I am not considering at this point. I therefore estimate a rise in total real agricultural wages within the economy of 14 per cent, which is equal to the increased number of such laborers as shown by the 1961 Census.

In 1950/1951 the Agricultural Labor Enquiry estimated the wage payment to agricultural labor at Rs 5.1 billion, or 10.5 per cent of the total income derived from agriculture in that year when it is valued at current

[21] Some of this may reflect the higher social status of "cultivators," and consequently the greater desire of respondents to say they were cultivators rather than laborers. But this desire should not have changed between the two census years, even though the ability to make the choice may have increased.

prices. Assuming the same proportions to real income expressed in 1948/1949 prices, the total real agricultural wage income in 1950/1951 equalled Rs 4.5 billion and the nonwage income Rs 38.9 billion. A rise of 14 per cent in the total real agricultural wage payments yields a figure of Rs 5.1 billion for real agricultural wages for 1960/1961. The nonwage agricultural income would then equal Rs 54.0 billion in 1960/1961, an increase of 39 per cent.

It is possible to make a crude distribution of the wage and nonwage income payments between the two classes of landholders below and above 15 acres. In 1950/1951 it is estimated that 60 per cent of the total payment of agricultural wages was made by the smaller landholders.[22] I assume the same proportion held in 1960/1961. On these assumptions Table 20 presents a breakdown of wage and nonwage incomes by size of holding.

TABLE 20

WAGE AND NON-WAGE INCOME BY SIZE OF HOLDING, 1950/1951 TO 1960/1961
(billions of rupees; 1948/1949 prices)

Size Holding	Agricultural Wages	Nonwage Income	Total	Per Cent Increase in Nonwage Income 1950/1951 to 1960/1961
1950/1951				
0–15 acres	2.7	24.3	27.0	
15+ acres	1.8	14.6	16.4	
	4.5	38.9	43.4	
1960/1961				
0–15 acres	3.1	32.0	35.1	+32
15+ acres	2.0	22.0	24.0	+46
	5.1	54.0	59.1	+36

Table 20 shows a proportionately greater increase in nonwage incomes going to the larger farmers on the basis of the previous, reasonably conservative estimates and assumptions; the rate of increase in such income for the larger farmers was almost 50 per cent higher than for the smaller ones. Although it would be very desirable to distinguish between rent and interest income and other income within the nonwage category, the data are lacking for any such distinction, even on the very crude basis of my earlier estimates. It is probable that the larger farmers receive a much

[22] D. Narain, *Distribution of the Marketed Surplus of Agricultural Produce by Size Holding in India, 1950–51*, p. 40, Table 4.

larger share of such income than the smaller farmers, but it is impossible to say what the overall trends have been with respect to these factor payments. The effects of the land reforms and rent ceilings probably reduced rental earnings of larger peasants and intermediaries. With respect to interest, coöperative credit has increased greatly, which would reduce private interest receipts; but much of this credit has gone to the larger farmers, and their loan activities may have increased at the expense of such activities by nonpeasant traders.

Although the larger peasants in general gained more than the smaller ones, the landholders of above 15 acres have been treated as a unified group. Certain qualifications should be entered. First, those peasants who were the largest before Independence probably lost somewhat as a result of the land reforms—abolition of intermediaries, land ceilings, and so on. The gains would thus be concentrated among the landholders of 15 to 30 acres. Second, in order to gain, the larger landholder must have the will to gain, the knowledge to take advantage of the government-sponsored innovations, and the capital resources to do so. Thus Brahmin or Rajput landowners who had never worked with their hands would not have known how to take advantage of the new developments on the land. The land-working as well as landowning castes—the Sikh Jats, the Gujerati Pattidars, and the Marathas—tended to be the gainers; and among these, gains would have gone to smaller as well as larger peasants in those castes. But the larger peasants in this latter group are the ones more likely to have the resources and the connections to take advantage of the changes. They would as a result be both more willing and more able to introduce new crops and techniques, especially those requiring a large investment, and they could best afford the risks involved in doing so.

Nonagricultural Groups in Rural India. The role of agriculture in the rural sector was still so dominant in 1960/1961 that changes in nonagricultural employment and income were relatively unimportant in the total. In my opinion the most significant change was the increase in the number of "other service" workers from 7.6 million in 1951 to 11.5 million in 1961. Unquestionably, a large proportion of these additional workers in rural "other services" were governmental or quasi-governmental employees. The community development, the village government, coöperation, the extension of banking facilities, education, health, and other programs all require large additions to government staff. The community development program alone has a central administration, a state administration, a district organization, and a project administration.

At every level there is an Executive Officer, functioning with the aid of a Development Committee and helped by an Advisory Board. At the Centre, there is an Administrator, at the State level there is a Development Commission, at the district level there is a District Development Officer . . . and at the Project level a Project Level Officer equipped with a staff of some 125 supervisors and Village level workers.[23]

With approximately 300 villages in a project area, there would thus be about 250,000 supervisors and village level workers for the 2,000 projects among India's 600,000 villages. When one adds the additional workers on higher levels of this one program, it would not be farfetched to estimate the employment of one community development staff member for each village. If to this one adds the other programs, an estimate of two additional government employees per village working in rural India as a result of the development effort would appear conservative. Such an increase would result in an increased employment in the rural "other services" employment category by more than 1.0 million people.[24] Apart from the direct economic and social effects of this additional group of employees and the programs they administer, they represent a new social force in rural India. They are on one hand a new group of intermediaries between the village and the government, replacing the traditional ones; on the other hand they are a source of power, influence, and patronage for the party controlling the government and for influential groups within the party.

Other categories of tertiary employment in the rural areas have increased by only a small absolute amount, and so have declined relatively. This may reflect the government's programs in the areas of coöperatives, banking, and trade, which may have reduced some of the business of the traditional moneylenders and traders both by directly replacing them with the new agencies or by making it possible for the larger peasants to handle parts of these activities independently.

[23] A. R. Desai, *Rural Sociology in India* (3rd ed.; Bombay: Vora and Co., 1961), pp. 548–550.

[24] This estimate agrees with one derived from simply looking at government employee statistics. It has been estimated that of the 6.6 million government employees at all levels in 1960, approximately 4.7 million were in urban areas and 2.0 million in rural areas. Applying the same proportions to the 3.5 million government employees in 1951 yields a figure of 1.0 million employees in rural areas and 2.5 in urban. Thus, the increase in rural government employment would have been 1.0 million in the period. J. M. Healey, "Public Employment in India, 1951–60," *The Economic Weekly*, XV, No. 23 (June 8, 1963), pp. 925–926; and National Council of Applied Economic Research, *Urban Income and Saving* (New Delhi: 1962), p. 31. This latter study gives a figure of 18 per cent of all urban households in 1960 deriving their income from government employment. I have applied this proportion to the figure of 26 million total urban employed in 1961 to get the estimate of 4.7 million urban government employees in 1960/1961. (An interesting report recently appeared in the Indian papers that a female and a male worker will be appointed by the Health Ministry to every village in India as family planning workers. See *Statesman*, June 22, 1963.)

With respect to employment in rural household industry and manufacturing, there has apparently been a large absolute increase from 6.7 million to 12.4 million workers in both groups; still this is a very small relative increase in employment. The data on this sector are most inadequate. In their study, P. N. Dhar and S. Sivasubramanian estimate an annual rate of increase in net output of approximately 8 per cent from 1955/1956 to 1959/1960, and approximately half of this would have been in rural India on the basis of their breakdown of the source of such income in the earlier year. The Khadi and Village Industries Commission cites high rates of increase of output by a small number of units for such rural home industry products as khadi (handmade cotton cloth), handwoven silk, wool, soap, matches, pottery, and vegetable oil. In many cases of handicraft industries, however, these increases have been subsidized, and there is a good deal of feeling that the rural industrialization programs have not been successful. The Third Plan, summarizing the report of the Evaluation Committee for Village Industries, stated, "that the results obtained in respect of both production and employment were not commensurate with the expenditure incurred," on various schemes in the Second Plan.[25] Since in the Second Plan these handicraft industries were to meet a large proportion of the additional domestic demand for consumer goods arising from the development program, as well as to employ rural population, their lag raises serious problems with respect to the economy's development. These industries also produce consumer goods that the peasant might wish to purchase, if price and quality were satisfactory, and thus are a mechanism for increasing the flow of the peasant's agricultural output to the market.

With respect to machine-using factory industries, there has probably been some increase in rural employment; but their relative importance was still minor in 1961, with less than 2 per cent of rural employment. Cement output, which grew rapidly in the decade, is usually located in India near raw material sources, and consequently its employment is in rural areas. There was also a rapid growth of coöperative sugar factories under the stimulus of both government coöperative programs and the pressure of land reforms. Some larger peasants converted their lands into sugar plantations or invested the proceeds of the sales of land into sugar factories that processed output produced by their lands. But these isolated factories apparently had little influence on the general pattern of rural society prior to 1951; and probably they had little general influence there-

[25] See Dhar and Sivasubramanian, "Small Enterprises . . . ,"; *Third Five-Year Plan*, p. 430. See also the criticisms as to both the concept and effectiveness of the present rural industrialization program by J. P. Narayan and Professor Gadgil, *The Hindu Weekly*, August 5, 1963.

after, although their local influence on the cropping and the sale habits of peasants and their impact by offering opportunities for employment to nearby villagers were more significant.[26]

Thus one can conclude that the expansion of the nonagricultural sector of rural employment played only a minor role during the past decade and would have only minor effect—even if it could be quantified—upon the distribution of the previous estimates of income gains in the rural sector. Whatever effects have taken place have probably supported the trends described with respect to agricultural incomes. The landowning peasants, who are often also members of the local dominant castes, have the knowledge of new opportunities and the resources to take advantage of them; and as well they have the political connections to get permits and so on. They have therefore probably made a substantial proportion of any private investment in rural industrial and commercial activities. These groups, because of their political influence and incomes and their access to education, would also be the most likely source of new government employees. This must be qualified by the reservation of a proportion of the government job opportunities for members of scheduled castes. If this general picture is correct, however, the net effect would be to increase the proportionate flow of rural incomes from sources other than agriculture to the higher income rural groups and the members of the dominant castes.

---

[26] See on this T. Scarlett Epstein, *Economic Development and Social Change in South India* (Bombay: Oxford University Press, 1962) in describing the effect of the construction of a cooperative sugar factory in Mysore upon neighboring villages.

9

Gains and Losses of Development—Urban India

Part II described and identified various components of an urban middle class and working class in India as of about 1951. What happened to the numbers of the urban employed or unemployed within both classes, and what happened to the real incomes of these classes both *in toto* and per capita from 1951 to 1961? The data with respect to members of these urban classes is even more fragmentary than with respect to the rural groups, and it will not be possible to present any consistent total income figures for the urban sector, divided into middle and working classes. At best it will be possible to indicate the trend of movement of the numbers of the identified groups within the two classes, and of the real incomes of these groups during the decade.

MIDDLE CLASS GROUPS

Table 21 is a summary table of the change in the numbers of the various identified occupational groups in the middle class from 1951 to 1961. The explanation of the individual figures is contained in Appendix C.

It would be ideal, if it were possible, to give a table presenting per capita incomes and income totals for the various urban middle-class groups for both 1951 and 1961. Unfortunately, aggregate statistics with respect to incomes by either factoral share or recipient group are simply lacking; and it is impossible to construct such a table with even the minimal accuracy used in the discussion of rural groups. However, it is possible, on the basis of government reports and other specific statistical data to present some indication of the absolute and relative trends of real incomes for the members of the various groups listed in Table 21, during the decade of 1951–1961, without adding up the specific parts to make an aggregate comparative statistical picture.[1]

[1] Almost no data exist for the self-employed, who are a large proportion of the urban labor force. These are probably owners of small-scale industries and retail trades (not peddlers) who run their own businesses with at most some family help, and whose incomes are a combination of wages, profits, and return on investment.

TABLE 21
CHANGE IN NUMBERS IN URBAN MIDDLE-CLASS OCCUPATIONS
(early 1950's to about 1961)

Occupational Group	Numbers of People in Labor Force (thousands)					
	Early 1950's[a]		1961[b]		Change in Period	Per Cent Change in Period
Employers	200		350		150	75.0
Merchants	2,300		3,000		700	30.4
Retailers		1,600				
Wholesalers		700				
Professional employees	770		1,220		450	74.0
Medical and health		230		300		
Teachers		390		710		
Journalists		100		150		
Lawyers, n.e.c.		50		60		
Administrators, executives and technicians (incl. clerks)	2,700		3,100		400	14.8
of which:						
Central government employees		300 (210)[c]		415 (290)		
Other government employees		425 (300)		1,400 (1,000)		
Manufacturing industry employees		160 (88)		210 (115)		
Bank and insurance employees		75 (53)		120 (85)		
	5,970		7,670		1,700	28.5
Educated unemployed:	200+		1,000+			

Notes:
[a] From Table 6.
[b] See Appendix C.
[c] Number of clerks in parentheses.

ENTREPRENEURIAL PROFITS

A first source for possible data with respect to profits is the income tax data. This, however, shows that the annual assessed incomes of companies increased by only 25–30 per cent in the period from 1951/1952 to 1960/1961. This small increase may reflect the changes in the legislation and the definitions with respect to both companies assessed and incomes, as well as possible underreporting; but in my opinion these results are much too conservative, even when one compares the tax data for company profits with Reserve Bank data over a similar period.[2]

The Reserve Bank of India has been measuring and analyzing the movement of company profits over the same decade. During the five years from 1950 through 1954 its survey included 750 public limited companies, that is, companies with a legal status similar to American corporations; and the coverage was raised to 1,001 companies in 1955 when comparable data became available for both the smaller and larger sample. Table 22 summarizes these data with respect to profit movements.

Table 22 shows an increase in the absolute level of money profits both before and after taxes on the order of 150 per cent over the eleven years. If averages for the first and last three-year periods are compared in order to eliminate short-term fluctuations, there are seen increases of approximately 100 per cent in total profits—before or after taxes—and in paid-out profits; meanwhile retained profits rose by only 70 per cent. If the increase in paid-out profits is deflated by the cost of living index, which rose by about 25 per cent over the period, the total real dividend income would have risen by about 75 per cent. If the increase in retained profits is deflated by the 27 per cent increase in the wholesale price index—excluding 1950/1951 when the index was very high because of the Korean War—real retained profits rose by about 45–50 per cent over this period.

The indicated rise in the absolute level of profits is supported by the Reserve Bank figures for the rate of growth of the companies included in its survey from 1950 to 1960. During the first five years, when only 750 companies were included, the average annual simple rate of growth of total net assets, when defined to include fixed assets net of depreciation plus circulating capital, was 7 per cent; for the 1001 companies the same simple rate of growth was 9.5 per cent in the second period from 1955 to 1960. The annual rate of return on capital—measured either as gross

[2] See Report of the Taxation Enquiry Commission, Vol. II, pp. 384–385; The Indian Merchants' Chamber, Executive Officer C. L. Gheevala, *National Income of . . . to 1960–61* (Bombay: IMC Research and Training Foundation, 1963), p. 245; Report of the Direct Taxes Administration Enquiry Commission, 1958/59 (New Delhi: 1960), Appendix XII. See Sources for Table 22.

TABLE 22
NET PROFITS OF PUBLIC LIMITED COMPANIES,[a] 1950–1960
(millions of rupees) [b]

Year [c]	Before Tax	Total	After Tax Retained	Paid Out
1950	640 (760) [d]	390 (450)	150	240 (290)
1951	850 (1,020) } (820)	520 (600) } (470)	280 } 200	270 (320) } (300)
1952	560 (670)	310 (360)	60	250 (300)
1953	650 (780)	390 (450)	140	260 (310)
1954	790 (950)	460 (530)	180	290 (350)
1955	1,190 (970)	680 (600)	290 (270)	390 (320)
1956	1,310	730	300	430
1957	1,090	570	140	430
1958	1,230	660	190	470
1959	1,630 } 1580	1,040 } 950	390 } 340	650 } 610
1960	1,880	1,160	450	710

Notes:
[a] This is net profits after depreciation.
[b] Rounded to the nearest tenth of a million.
[c] The year is July 1 to June 30th; that is, 1950 would run from July 1, 1949 to June 30, 1950. 1950 through 1954 includes 750 companies; 1955 through 1960 includes 1001 companies.
[d] The figures in parentheses are the estimates for the 1,001 companies for the earlier year, which were reached by raising the estimates for the 750 companies for the earlier year by the percentage difference between the figures for the 750 and 1,001 companies. In 1955 the figures for both groups were available. The retained profits figures are the actual ones; there seems to be little difference between the two 1955 figures. The braced figures are drawn to three-year averages.

SOURCE:
Reserve Bank of India Bulletins: "Finances of Indian Joint Stock Companies 1955," September, 1957, "Finances of Indian Joint Stock Companies 1956," October, 1958; "Finances of Indian Joint Stock Companies 1960," June 1962; "Survey of Ownership of Shares in Joint Stock Companies as at the End of December 1959," May, 1962 (Bombay: Reserve Bank of India).

profits before tax defined to include interest charges, managing agents' remunerations, and tax provisions as a percentage of total net assets, or as profits after tax as a per cent of net worth—remained at an average of about 9 per cent per year during the entire period. The annual rate of growth resulted roughly in a doubling of the total net assets of the companies over the period; when this is combined with a constant rate of return on capital, the effect is an approximate doubling in the absolute level of profits that was observed.

From various surveys of stock ownership and dividend income, it appears that there is a high concentration of ownership in India. In 1954/1955, of a surveyed group of individuals paying income tax, 50 per cent of those surveyed who received dividend income had incomes below Rs 4,000 per year, but this group received only about 10 per cent of the dividend income; at the other extreme the 1 per cent of those who received dividend income and had incomes in excess of Rs 100,000 per year received over 30 per cent of the dividend income. A 1959 survey of stock ownership of a selected group of companies showed that individuals held approximately 99 per cent of the accounts. Of these individual stockholdings, approximately 88 per cent were valued below Rs 1,000, but this group of small owners held only 14 per cent of the total value of stock held by individuals; at the other extreme those individuals in the large holding group of Rs 50,001 and above had less than 1 per cent of the total individual accounts but owned 20 per cent of the total value of individually held stock. Both these surveys indicate that nearly all the dividend income in India goes to the small proportion of the population paying income tax; and within the taxpaying group the largest proportion goes to those in the higher tax brackets, and thus should be reflected in their incomes. This is probably not peculiar to India, but exists in the United States and other private enterprise economies as well.

I should like to add that there is believed to be substantial underreporting of profits for taxation purposes. Nicholas Kaldor estimated this as high as Rs 2.0 billion in mining and factory establishments alone in 1953/1954; and although this has been criticized as being too high, a government commission gave examples of discovered concealed incomes in all areas in one year totaling Rs 1.5 billion. Much of this would probably represent profit income since salaries would be public and declared. The effect of any underreporting would be to underestimate the absolute if not the proportionate increase in entrepreneurial incomes during the decade. I shall not attempt to make any estimate of changes in per capita or per firm profit income, which would be impossible from the data and would signify little. However, if these estimates are correct, it is possible to conclude that industrial profit income was rising at a significantly faster rate than either real national income or total urban income; and thus it is clear that the share of such entrepreneurial income to total income was also increasing, and much of this increase would be going to the urban higher-income recipients.[3]

SALARIES OF SENIOR EXECUTIVES AND EARNINGS OF HIGH INCOME PROFESSIONALS

In Chapter 2, I estimated that in the early 1950's the number of individuals in the urban upper middle class, which consists of those individuals assessed for income tax on earned annual incomes of Rs 12,500 and more, was about 70,000; and in the urban middle middle class, those in the Rs 3,600 to 12,500 bracket, the number was 300,000. In 1960/1961 the number of individuals in the former class had risen to 141,000; and in the latter, which included by then all those earning from a Rs 3,000 annual income base—since the income minimum for tax payment had been reduced—the number of individuals was 673,000. Thus, the number in both these groups had more than doubled. At the same time the total income of those individuals assessed to tax rose from approximately Rs 2.0 billion in 1952/1953 to almost Rs 8.0 billion.[4] This would result in a significantly higher figure, at the minimum, for the real income per assessable individual during this period—allowing for a 25 per cent rise in cost of living.

Within the population paying income tax a major segment is the salaried income group, which includes a large proportion of both professional and administrative employees in the middle class. Since a very detailed government report has been issued on the comparative position of the members of this group employed by the government during the period from 1948 to 1957 most of the discussion will be based upon this report, supplemented by information for 1960/1961. Table 23 presents data in individual and in total terms with respect to salary movements of those income tax assessees employed in the government and private sectors since 1948/1949.

The total number of salaried income tax assessees approximately doubled from 1948/1949 to 1960/1961; over the same period the total

[3] Nicholas Kaldor, *Indian Tax Reform—Report of a Survey* (New Delhi; GOI Ministry of Finance, Department of Economic Affairs, July, 1956), pp. 103–106; *Report of The Taxation Enquiry Commission,* p. 148, on income tax evasion. I believe this figure of Rs 1.5 billion is in a single year, but this is unclear.

[4] I am using approximate figures because the income tax minima were changed during this period, and also because I do not have tax data for 1952/1953 when I have the above income class breakdowns. Total income assessed to all assessable individuals in 1951/1952 was Rs 1.6 billion; for 1953/1954 it was Rs 2.1 billion; for 1960/1961 it was Rs 7.8 billion.

salary income received by these assessees rose by over 150 per cent. During this period the average pre-tax and post-tax salary income per assessee rose by about one-third—even so, it fell from 1956/1957 to 1960/1961—so that the real salary income per capita for the greater number remained roughly constant. With respect to the group with salaries over Rs 40,000 the number of assessees increased by over twelve times since 1948/1949, but the average individual pre-tax money salary fell by about 10–15 per cent and post-tax salary by over 25 per cent. In real terms the decline in per capita real salary income was much greater since the cost of living rose by about 25 per cent during this period.[5] It is probable, however, that the decline in the average of money and real incomes does not reflect any actual declines in incomes by those members of this group who were members of it in 1948/1949, except possibly government employees. Rather, as more people entered the group earning Rs 40,000 and above the newcomers would cluster at the lower level, say Rs 40,000 to 50,000 per year, thus reducing the average salary for the group.

Within the period from 1948/1949 to 1956/1957 it is possible to compare salary movements between government and private employees. During this shorter period, as the cost of living index rose by only 7 per cent, there was a gain of only 5–10 per cent in real salary income per salaried government employee included in the previous table compared with private employees. In fact since there had been no increase in salaries for government workers earning above Rs 300 per month from 1947 to 1957, any increase shown in either the numbers or the real salary of the tax assessable salary recipients in the government would largely reflect promotions or filling of vacancies in government offices.

With respect to the highest income levels in government during this period from 1948/1949 to 1956/1957, there was a sharp decline in real income compared with a relatively constant real income in the private sector. Since the Commission of Enquiry recommended against any increase in salary for those government employees earning above Rs 2,000 per month, the position of this highest salaried group has deteriorated even further, both absolutely and relatively, since 1957, though the cost of living index rose by about 15 per cent from 1957 to 1961 alone. Thus for this group also there has been a sharp deterioration in relative salary income compared with those in the private sector. However, it is not possible to conclude from this that any one employee in this group of government executives in 1960/1961 is now in a worse position than he was in

[5] This does not say that the total real per capita income from all sources of this group fell, since salary income was less than 30 per cent of the total income of the group receiving total incomes above Rs 40,000 per year in 1960/1961.

TABLE 23

MOVEMENT OF SALARIES OF SALARIED INCOME TAX ASSESSEES, 1948/1949 to 1960/1961

Class of Salaried Earners	All Tax Assessees			
	1948/1949	1956/1957	1948/1949 to 1956/1957 (per cent change)	1960/1961
Numbers [b]				
In government	59,845 *	47,404	−20	
In private business	136,559	159,145	+17	
Total	196,404	206,549	+ 5	382,259
Salary incomes before tax				
Total (millions of rupees)	1,130.4	1,802.1		2,910.8
Salary income per assessee (rupees)				
Government	6,567	7,426	+13	
Private	5,400	9,112	+69	
Average	5,755	8,724	+52	7,608
Salary income after tax				
Per assessee (rupees)				
Government	6,096	7,000	+15	
Private	4,795	7,950	+66	
Average	5,191	7,731	+49	6,838

TABLE 23 (Con't.)

MOVEMENT OF SALARIES OF SALARIED INCOME TAX ASSESSEES, 1948/1949 to 1960/1961

Class of Salaried Earners	Assessees with Salaries Above Rs 40,000 per Year [a]			
	1948/1949	1956/1957	1948/1949 to 1956/1957 (per cent change)	1960/1961
Numbers [b]				
In government	180 [c**]	270 [c]	+ 50	
In private business	321	1,970	+514	
Total	501	2,240	+471	6,869
Salary incomes before tax				
Total (millions of rupees)	28.4	136.1		336.2
Salary income per assessee (rupees)				
Government	55,000 [c]	42,000 [c]	−25	
Private	57,477	63,356	+10	
Average	56,687	60,759	+ 7	48,994
Salary income after tax				
Per assessee (rupees)				
Government	38,000 [c]	32,000 [c]	−15	
Private	38,629	37,367	− 3	
Average	38,324	36,694	− 4	27,515

* Central government only.

** Central and state governments.

Notes:

[a] There were two people with salaries above Rs 200,000 in 1948/49 and twenty-nine people in 1956/1957. All were in the private sector.

[b] The exemption limit in 1948/1949 was Rs 3,000; in 1956/1957 it was raised to Rs 4,200, and in 1957/1958 it was again lowered to Rs 3,000. In 1957/1958 the number of assessees employed by the central government was 49,439; by private industry 171,130; and by both 218,534.

[c] These figures for government employment and incomes are estimates from the text surrounding the table on p. 82 of the source. I have taken the mean average of the Rs 40,000–70,000 range for 1948; for 1956/1957 I have reduced this by 25 per cent. These adjustments are explained on pp. 81 and 82 of the source. The post-tax government estimates are rough guesses based on this movement, and utilizing the proportions in the source table as a guide.

SOURCES:

GOI Ministry of Finance, Commission of Enquiry on Emoluments and Conditions of Service of Central Government Employees, 1957–1959, *Report* (n.d.), pp. 81–84; Indian Merchants Chamber, p. 250, Table 13.15 for 1960/1961 data.

1948/1949. Many employees have been promoted into the high salaried group; others have retired from the government, some to higher private salaries. The effect is rather that private salaried employment has become a good deal more attractive than it was earlier when compared with what is offered by government employment in thc higher salary ranges.[6]

In conclusion, it is possible to say that the group of salaried employees paying income tax benefited by the decade of development. With the great increase in the number of such employees the total of salary income earned by this salaried group rose from approximately 1 per cent of the money national income in 1948/1949 to 2 per cent in 1960/1961. The apparent decline in per capita salaries from 1956/1957 was most likely not an actual decline, but a reflection of the great increase in the number of salaried income tax assessees since that date; furthermore, it may be more than compensated for by other types of income gains. However, almost all the per capita money income gains have been achieved in private industry; within the government the higher salaried income groups have lost in economic terms, both in comparison with the higher salaried groups in private industry and with the lower salaried groups within the government itself. With respect to the latter the high-salaried government workers still have many perquisites that the low-salaried ones do not have; furthermore, some of these losses may be made up over an individual's lifetime by the greater opportunities for subsidiary income and for transfer to private industry at much higher salaries upon their retirement from government. For example, there are also greater possibilities for members of a government officials' family to get private employment at relatively high pay; and there may be other semi-legitimate sources of personal income as private businessmen seek access to permit-granting government officials.[7] But these subsidiary legitimate economic benefits in India are probably still of minor value in all. The total of these indirect benefits, in my opinion, would far from recompense in money terms the real degeneration in the position of senior government officials.

With respect to the professional recipients of higher incomes—doctors, lawyers, screen stars, and the like—statistics have almost nothing to offer. The income tax statistics show a high degree of constancy in both the number of professional assessees and the income assessed to taxation. It is believed, however, that there has been a great deal of underreporting

[6] In both the government and private sectors the highest salary groups have many perquisites—housing, transportation, retirement benefits—to make up for some of the apparent decline in average real salary. I cannot compare the private and public employees with respect to these over time.

[7] There is frequent criticism of these side income possibilities in the Indian press, and much discussion of the "retirement" problem, and the "relatives" problem, as opening possibilities of unethical influence.

of income by this group. Kaldor estimates that less than 25 per cent of the professional income above the exemption limit is assessed. There are many rumors of cash under-the-table payments to cinema stars and doctors and lawyers of highest income. It would be expected that the demand for such professional service has a high income elasticity; and with the great increase in the number of high-income managerial and administrative personnel and the total income of this group, that professional incomes should increase in a roughly comparable fashion. I believe they have over the decade—probably at a cost of leisure time—regardless of the tax data.

RETAIL AND WHOLESALE TRADE PERSONNEL (LARGELY SELF-EMPLOYED)

Data with respect to incomes of this group are quite inadequate, and the official statistics probably underestimate the rise in income from the commerce and transport sector.[8] In money terms the official statistics show a rise of about 30 per cent in the total national income derived from the "other commerce and transport" sector during the decade. However, the real national income from commerce, transport, and communications—which includes the smaller group—rose by about 50 per cent; and in the earlier table of rural and urban income changes, I estimated a 45 per cent gain in real urban income from this entire sector during the decade. I use this last estimate as my basis for further analysis. The number of people classified as retailers and wholesalers—exclusive of hawkers, peddlers, and the like—rose by only about 10 per cent over this same period. If it is assumed that approximately one-half of the total 45 per cent increase in real urban income over the decade from the commerce, transport, and communications sector has gone to retailers and wholesalers, this would result in a 20–25 per cent increase in this group's total real income and thus an increase in the real income per retailer and wholesaler of about 10–15 per cent.

LOWER MIDDLE-CLASS INCOMES

Government Employees. This group consists of clerical workers—not sweepers, bearers, and so on—employed by the various governments at salaries below Rs 300 per month. Roughly, they would be in Class III of the government hierarchy. Although the Report of the Commission of Enquiry into central government pay scales contains data for salary movements of central government employees, little if any data are available with respect to salaries of state and local government employees; but I think it reasonable to assume that salaries for state and local government

[8] K. N. Raj, in his frequently cited article of February 4, 1961, in *The Economic Weekly*, has an excellent analysis of the vagaries of the estimate of "other commerce and transport" incomes.

clerks have not moved more favorably than those for central government clerks.

In 1957/1958, at the time of the report, the minimum clerical salary in the central government was Rs 115 per month. The commission estimated that the real salary income of this group may have declined slightly from 1946/1947, and increased slightly compared with 1947/1948. I think it would be correct to say that real incomes of these low-paid clerical workers, in the Rs 115–300 per month range, did not increase during the ten years prior to the report; and the effect of the commission's recommendations with respect to their salaries would simply keep their real incomes roughly constant. The commission also made some comments on the relative salaries of these clerical workers compared with other groups in society over time. In 1947 the differential between the salary of a low-level government clerk and the wage of a government manual worker was 64 per cent; by 1957/1958 this had fallen to 53 per cent. An earlier commission had estimated the cost of living of a lower-middle-class family at about 80 per cent above that of a working-class family.[9] Although the statistical support for this cost of living estimate is weak, if it is accepted as being roughly accurate, this estimate and the previous data clearly show, first, that the economic position of lower-middle-class employees of the government has been deteriorating relative to factory laborers as well as to higher-middle-class families employed in private industry; and second, that the effort on the part of these lower-middle-class families to maintain a traditional difference in status from working class families and to "keep up with the Joneses" has become more difficult.

Private Clerical Employees and Others. Some data are available with respect to the movement of bank clerks' salaries during the decade. Bank clerks are, in my opinion, reasonably representative of the urban lower middle class in terms of living conditions, and they may be in a somewhat favorable position with respect to their strength as a group because they are strongly organized. In 1962 the range of pay for clerks and subordinates in banks was a minimum of about Rs 75 per month to a maximum of Rs 400. This included base pay plus a cost of living allowance, and varied with the size of the bank and its location and with the seniority and service time of the individual. This scale was the result of a series of investigations and government awards going back to 1949. In 1962 another tribunal recommended an upward movement in the salary scale of bank clerks from the previous one to a range with a minimum of Rs 104 per month, and a maximum of Rs 405 per month, both of which include cost of living allowances. The tribunal based this decision on a finding that

[9] *Commission of Enquiry on Emoluments Report,* chap. 7, p. 105.

there had been some deterioration in the real earnings of bank clerks from 1949 to 1962, and the effect of the recommended pay increase would be roughly to regain the 1949 position.[10] On the basis of this it is possible to conclude that the position of bank clerks—and probably insurance agents, store clerks, and such—has at best remained constant during the Plan periods.

With respect to clerks in manufacturing, far fewer statistics are available. Certainly those clerks now working for newer, rather high-paying, foreign-owned companies, such as the oil companies, have probably improved their position; but clerks in Indian-owned private firms probably have done little if any better than those in the government or in banks. In the period from 1948 to 1958 in the jute industry of West Bengal the salary of a clerk with an equivalent of a high school education had fallen from 75 per cent to only 53 per cent above that of an unskilled manual worker; in the engineering industry in the same state, the same differential had fallen from 83 per cent to 58 per cent; in the Bombay textile industry it had declined slightly from 46 to 44 per cent.

Data with respect to the movement of teachers' salaries are unavailable, but data are available with respect to the structure of university and college teachers' salaries in 1957/1958. Of 36,000 full-time university teachers, approximately 70 per cent were receiving incomes below Rs 301 per month, and less than 10 per cent received incomes above Rs 450 per month.[11] The salary scales of teachers below the college level are far lower. Relatively, compared with government administration employees in similar fields, teachers' salaries are probably lower. It is likely that the real incomes of teachers have not increased significantly over the decade.

General Conclusions on Lower Middle Class Income Movements. A recent survey of the incomes of middle-class families in the four major cities of India—Calcutta, Bombay, Delhi, and Madras—showed that in 1958/1959 the average monthly income of all the families ranged from about Rs 300 to about Rs 380 per month, of which Calcutta had the highest average and Madras the lowest. About 60 per cent of the families had incomes below the average of Rs 300 per month, with the maximum concentration in the range of Rs 100–200 per month. I think it a fair summary of the changing position of the lower middle class—defined as white-collar employees with incomes below Rs 300 per month—during the development decade, to say that although the numbers in this group

---

[10] GOI National Industrial Tribunal (Bank Disputes), *Award on the Industrial Disputes Between Certain Banking Companies and Corporations and Their Workmen by Presiding Officer* (New Delhi; June, 1962), especially chap. 5.

[11] GOI Ministry of Education, *Education in Universities in India, 1957–58* (New Delhi: 1961), p. 22. See also D. D. Karve, "Emoluments—Administrator and Educator," *Opinion,* June 18, 1963, pp. 4–6.

have increased significantly, probably by over 100 per cent in the decade from 1951 to 1961, the real income per employee has probably remained roughly constant. Thus the total real income of this class has also about doubled, in proportion to its numbers.

This doubling of the numbers of the lower-middle-class members represents a gain to Indian society and economy which arises from the development effort. These same employees, without this expansion in number, would probably have been in low-paying employment; or they would have been unemployed, in either a disguised or open form, if growth had not occurred. However, it is possible that there is now a wider gap between the expectations of this larger group and their accomplishments than there was previously, since many have not received the great opportunities they thought their education opened to them; and this is a source of potential political unrest. This gap would, naturally, be greatest among the educated unemployed.

The threat of political unrest from the lower middle class is probably increased by the relative deterioration of its economic position in comparison with that of both the upper income groups and the factory workers in private industry. Even the more powerful classes in government at least now have much greater power if not higher income. The strain of attempting to maintain former positions relative to classes felt to be lower or to others of similar class background who have risen more rapidly is a potent source of unrest and dissatisfaction. As happened in West Bengal, this may lead some members of the lower middle class to demand revolutionary political change of a Utopian character and to support parties with such programs; and when the class position is associated with caste and communal characteristics, as it frequently is, it may lead to strengthened efforts to retain caste or communal characteristics and positions and, thus, to support of communal parties such as the Jan Sangh in the North and the DMK in the South.[12]

WORKING-CLASS GROUPS

Table 24 presents in summary an estimate of the change in the numbers in various groups of the urban working class, exclusive of peasants, in 1951 and 1961. The basis for these figures for 1961 are given in Appendix D.

FACTORY AND GOVERNMENT MANUAL WORKERS

This is probably the most important as well as the most easily identified group within the working class. Unfortunately the statistics with respect to

[12] This is an explanation given for the volatility of the Calcutta white-collar electorate.

TABLE 24
BREAKDOWN OF URBAN WORKING-CLASS OCCUPATIONS, 1950's to 1961
(thousands)

Occupational Group	Early 1950's	About 1961	Change	Per Cent Change in Period
Operatives and artisans	3,500 (1,800)[a]	5,100 (2,700)[a]	1,600	45.7
Household industry	2,040	2,100	60	2.9
Service workers	3,100	3,400	300	9.7
Hawkers, peddlers, etc.	550	900	350	63.6
Building industry	400	500	100	25.0
Unskilled	1,860	2,800	940	50.5
Total	11,450	14,800	3,350	29.3
Unemployed	about 500	about 1,500		

Note:
[a] Employees in registered factories. C. A. Myers, *Labor Problems,* gives a total registered factory figure of 2.6 million in 1954; the *Indian Labor Journal* (January, 1964) gives a figure of 3.9 million in 1961. It is estimated that 70 per cent of employees in registered factories live in urban areas.
SOURCES:
Table 7 for 1950's. Appendix D for 1961.

the movement of real wages of factory labor since 1955 are inadequate. Before 1956, S. A. Palekar has shown from a detailed analysis of industrial-wage and cost of living data, that real wages in Indian factory industries rose by about 30 per cent from 1950 to 1955. The Labor Bureau of the Government has estimated that the real earnings of persons employed in factories and earning below Rs 200 per month—which would include not only most operatives but also some clerks and service workers, a group that is somewhat different from Palekar's—rose by 24 per cent from 1947 to 1959, but only by about 9 per cent from 1950 to 1959. This rise was concentrated during the First Plan and thus agrees in general with Palekar's results for this period; from 1955 to 1959 the real earnings of this income group of factory employees declined by approximately 15 per cent.[13]

I think there was some increase in real wages of factory workers during this decade. Especially in the early part of the decade, the government was strongly in favor of raising worker wagers on ethical and health grounds. Many of the factory workers are in trade-unions that are affiliated with the Congress party, and the government has sought to assist these unions in their conflict with unions not so affiliated. The period of the First Five Year Plan was relatively successful with several very favor-

[13] S. A. Palekar, *Problems of Wage Policy for Economic Development* (Bombay: Asia Publishing House, 1962), chap. iii; Government of India, Ministry of Labour and Employment, Labour Bureau, *Indian Labour Statistics 1961* (Simla: 1961), pp. 19–26, 107. See on this "Trends in Money and Real Wages in India: 1951–61," *Reserve Bank of India Bulletin* (April, 1964), pp. 421–439.

able crops, and it was characterized by expanding use of idle capacity so that prices declined somewhat, and there was no great pressure for austerity or "holding the wage line." The system of factory wages, which often includes a cost of living allowance, which is adjusted to changes in the price index as a component of the wage, together with the system of arbitration of disputes, which tends to compromise workers' demands, contributed to a substantial improvement in real wages during the First Plan. In the Second Plan the government's wage policy changed, since its expanded development program called for more resources and for "holding the wage line," especially in the potential export industries. At the same time, under the pressure of increased development expenditures and higher excise taxes the cost of living rose by approximately 5 per cent per year after 1955/1956. Although for many of the factory workers the price increase has been counterbalanced by cost of living allowances within wages, approximately 60 per cent of the total number of workers were not covered by such contract provisions. However, in overall terms and in spite of the decline in per worker real wages from 1955 to 1961, the economic position of the group of urban factory workers improved during the decade in terms of numbers employed and total real wages.

The government also employs a large number of manual workers on railroads and similar activities. The movement of wages of this group of workers has been roughly the same as for lower grade clerks, and the effect of the recommendations of the Pay Commission has been to maintain real wages. These wages are also maintained at a level comparable to those of similar workers in private industry, to discourage raiding.

For private and government factory workers there has been a definite increase in nonwage benefits. An employer-financed provident-fund system has been introduced; also many of the plants, especially the newer ones, have introduced subsidized food and housing programs, and these programs have resulted in a gain of real income if not in money income. It is estimated by an employer group that the value of these nonmoney income benefits of private industrial workers was the equivalent of about 12 per cent of total annual money earnings in 1960.[14]

OTHER WORKERS

There are few statistics for the wages of the 8–10 million workers in urban unorganized activities—those in household industries, services, hawking and peddling, building, and so on. In qualitative terms this is the weakest group of workers in India: they are largely unskilled, unedu-

---

[14] The Employers' Federation of India, *The Wages of Industrial Labour Since Independence*, Monograph No. 3 (Bombay: 1963), pp. 18–28.

cated, and unorganized; many are marginal workers, on the fringe of the urban areas and moving between rural and urban work. In part because they are on the fringe they accept low wages and casual urban employment since much of what they earn is sent back to their families in the villages. Although the average real wages for workers in this marginal group may show a fall during the decade as a result, the urban wages earned are not comparable to those in regular factory or commercial employment. Rather they represent part of an increasing total of real money earnings by rural families, but at a heavy social cost of family separation and squalor.

The only independent statistic of wage movements possibly related to this group is one given in Palekar which shows a decline of 30 per cent from 1950 to 1955 in the real wages of workers employed in personal service industries—laundry, dyeing, and cleaning.[15] It is probable that a better measure of wage movement for this group is the movement of real earnings of agricultural labor, since these workers are on the fringe of both urban and rural occupations; and the earnings of farm laborers have at most declined by about 10 per cent, and at best remained roughly constant over the decade.

SUMMARY OF SHIFTS IN ECONOMIC POSITION OF VARIOUS URBAN CLASSES AND GROUPS

From the previous discussion, even granting the very poor statistics, I think that with one major exception almost all class groups within urban India have shared in the absolute gains of development during the past decade, either by greater employment or by higher per capita real incomes, but in only a few cases by both. However, some major groups have lost relative to other groups.

The two groups that have probably gained the most are, first, the high-income members of the middle class in the private sector, defined broadly as those earning above Rs 3,000 per year or more narrowly as those earning above Rs 12,500 per year; second, the organized factory workers, especially those in the private sector, though their gains were confined to the period from 1950 to 1955. Both groups would have gained by the industrial growth that has occurred, for they have the skills or resources essential for that growth.

The one group that has clearly lost, relatively and absolutely in terms of the real income of its members, is the higher income government

[15] S. A. Palekar, *Problems of Wage Policy for Economic Development*,

salaried personnel—probably all those in the income-tax-paying brackets, and certainly those in the highest salary group. This is not inconsistent with gains by all the present members of this group, if they have been promoted significantly since 1947, as many have been.

The members of the lower middle class, the below Rs 300 per month group, have probably been fortunate to maintain their real income level during this decade; and they have probably lost in relative terms both to the high-income middle class, whom they try to emulate, and to the organized wage earners in factories, whom they struggle to keep ahead of. The numbers of lower-middle-class employees have increased greatly, however. Similarly, I think members of the unorganized and unskilled working class have at best barely maintained their level of real income, but the additional incomes earned by some of the individual members of this group probably have added to the incomes of the rural families that many of these unorganized urban workers continue to support.

In spite of the increase in numbers of employed, the numbers of openly unemployed—both educated and uneducated—have also increased greatly in urban India, approximately tripling during the decade.

10

Caste and Communal Economic Gains and Losses Since Independence

In previous chapters, I measured the economic gains and losses of various identified class groups within Indian rural and urban society. There are at least some statistical data available to make such an effort feasible. I am now going to attempt a somewhat similar effort with respect to the various caste and communal groups within Indian society—ethnic groups that cut across the economic class lines. Unfortunately there are almost no statistics on caste incomes or powers. Furthermore, the changes in caste economic position are so closely related to shifts in political power among the castes that it is difficult, if not impossible, to unravel the connections. Much of the discussion in the previous chapters on shifts in political power during the past decade is therefore closely related to the economic presentation in this chapter.[1]

SHIFTS IN ECONOMIC POWER IN RURAL AREAS

In the discussion of caste in rural India, the major theme was the position of the dominant caste within a village, and its traditional role within the economic system of the village. I pointed out that the dominant caste is the center about which patron-client relations within a village revolve. I also stressed that although the characteristics of the patron-client system and the position of the dominant caste are similar in many respects throughout India, the specific caste or communal group that plays this role varies from region to region in India. This, of course, is one of the major difficulties in measuring shifts in economic power on any national basis. The major characteristic of traditional rural caste dominance is landholding. Around these landowning castes there are the Brahmins, with their high ritual status; various service castes such as

[1] The only general statistical data, with respect to shifts in communal economic power, that might be relevant are the very poor statistics of state incomes and outputs. To a certain degree the states have represented people of the same community; and since the reorganization in 1956 which based the state borders on linguistic grounds, this is even more so. There is also a tendency for the same linguistic groups to have similar caste structure. These state income data are presented in Appendix E. Needless to say, this will tell us nothing of intercaste variations within a state.

traders and moneylenders, carpenters, smiths, pottery makers, barbers, and so on; and finally the group of so-called untouchables, who are largely included within the scheduled castes. As I pointed out, within the traditional system there was a possibility for all except the unclean castes to achieve either dominance or some of the major attributes of dominance with respect to economic, social, or political power. I will try to judge the improvement in economic position of these nondominant service castes, as well as the dominant caste.

In general the non-Brahmin dominant castes have probably gained economic power. In the analysis of class economic gains in rural India I pointed out that the larger landholders, with fifteen acres or more, who have the experience and knowledge to gain, have probably gained. They have gained as landholders, since they have been the ones best able to take advantage of the flows of resources and additional technical knowledge into the village during the two Five Year Plans. I think it is a valid generalization that the members of the dominant castes are likely to be the larger landholders in a village, and these gains would thus flow to the dominant caste.

In earlier chapters I pointed out that among the first results of Independence was the elimination of the political power of the princes in the former princely states, and furthermore that probably the major land reform in rural India was the elimination of the intermediaries. The effects of this change—especially upon the larger zamindars—combined with steps being taken to limit the size of holdings, have probably weakened the economic position of previously dominant groups. Thus, those who were formerly the largest landholders have lost in terms of both economic and political power; this is supported by the Reserve Bank data with respect to changes in income distribution. Meanwhile the groups of landholders below the largest—say the group with 15–30 acres—have probably gained relatively in both these closely related aspects of dominance.

I have already mentioned several recent specific studies [2] of local politics, in which one of the major trends noted was the widening of political power from numerically very small communal groups, such as the largest zamindars and princes, to somewhat larger numerical groups, but still landholders. These studies also point out that the political leaders within the district are also chairmen and leaders of the new economic institutions set up during the Plans. Thus the members of politically dominant castes within an area are also those best able to take advantage of

[2] See especially the mimeographed studies of Madurai and Belgaum politics by M. Weiner; see also L. Shrader and R. Joshi, "Zilla Parishad Elections in Maharashtra and the District Political Elite," p. 151.

the new institutions and of their resources; at the same time there has been some change in the position of formerly dominant castes to give greater weight to those landholding castes with larger numbers.

One other important factor influencing the distribution of gains among the farming castes is the one stressed by Kusum Nair in *Blossoms in the Dust:*

> The best farmers are to be found not necessarily in communities most favorably endowed with material resources, but in those that are traditionally agriculturist by caste. . . . It is so mainly because these castes have an inherited respect for agricultural work and they are not precluded by religion or tradition from working on the land. The result is that . . . members of these "professionally agricultural" communities . . . will be found superior in husbandry to the non-working castes of landowners, such as the *Brahmins, Rajputs, Banias,* etc., though the latter may have more capital, land, education, and thus superior means to acquire modern techniques and tools of cultivation. . . . [There] is [also] no uniformity yet in the prevailing value systems which determine not only a community's pattern of production and consumption, of farm management . . . but also its primary attitudes and wants. These vary greatly from one community to the next, within groups in the same region and even locality. . . . [Thus] a uniform response to common incentives and stimuli cannot be expected.[3]

The entire set of conclusions need not be accepted;[4] but it appears reasonable that those castes and communities with greater tradition and experience in farming their own land will be better able to take advantage of incentives and knowledge. I should also expect that within these communities and castes those peasants with greater resources could better do so than those with fewer resources. Where the experienced landworking castes are also the dominant castes, their gains from the development programs would be greatest; where those castes are not dominant, their gains in absolute and relative terms would be limited by their relatively limited resources. This conclusion is at least partly supported by a comparison of the agricultural development in other states with development in Gujerat, Rajasthan, and Punjab, in many areas of which castes with great experience in landworking are dominant. (See Appendix E.)

If the dominant castes have gained in rural India, what about the Brahmins and other "clean," though possibly not dominant, castes? I have already stressed that under the British the Brahmins had in many areas

---

[3] Kusum Nair, *Blossoms in the Dust* (London: Gerald Duckworth & Co., Ltd., 1961), pp. 190–191.

[4] Very many village studies and several statistical studies of crop shifts do indicate a high degree of price responsiveness and awareness by Indian peasants; it is possible to argue that economic motivations have in fact not been given enough scope in economic policy. See chap. 11 below.

sold their land and had moved to urban areas and taken up administrative, intellectual, and other white-collar positions. In some rural areas, such as Kerala, however, they kept their lands and were dominant. I think that in both the rural and urban areas the Brahmins are relatively small in number. In Chapter 2, I presented statistics on the caste structure of rural India showing that less than 10 per cent of the Hindu rural households were from the upper castes, which includes other castes besides Brahmins. In a democracy with universal suffrage, this would mean that these caste groups would either lose political power or have to share it with other groups to retain their political position. This loss in power has in fact happened, and it is reflected in shifts in the rural position of the Brahmins as well as their urban position. Reservation of seats in schools and positions in government for scheduled castes has cost the Brahmins the most, since they had been in the past the ones who were more likely to enter school and government; and it has affected rural as well as urban Brahmins. Since they are rarely dominant in terms of number or in landholding, they would also gain relatively less from the new rural institutions set up by the government and controlled by politically dominant groups.

The Brahmins are also affected because—as pointed out by Kusum Nair—even if they still retain their land, from experience and tradition they are generally less able to take advantage of the new opportunities. They are more likely to be landowners than land operators; they are often inhibited from operation of their land by their caste restrictions. Thus other caste groups would be able to gain relatively more from the developments that have occurred in agriculture.

It is much more difficult to speak of other nondominant, clean castes that traditionally performed semiofficial intermediary functions—that is, they assisted in the village government and represented the village in contact with the outside world. Patwaris, kulkarnis, and gowdas kept the village records and assisted the village panchayat leaders in government of the village. They might be members of the dominant caste, or they might be, and frequently were, members of client castes. These caste groups probably lost in both economic and political terms. As a result of the community development and village government reforms, new officials were brought in either to perform similar functions or to supervise the traditional intermediaries. In any case the newer officials have superior intermediary positions in the sense of more direct contacts with the central and state governments; formerly, too, the members of the intermediary castes received not only direct income from their work, but they got additional incomes via bribes for such acts as adjusting village records. Such payments are now more difficult to receive and are possibly lower, since there are more people involved. Thus the individuals performing these traditional intermediary functions have probably lost in terms of status, power, and money. At the same time, quite apart from the new government officials, the effect of the greater integration of the village into the larger political and economic area has brought into existence a new group of informal intermediaries; in many cases, these are businessmen who live sometime in the village and sometime in the city, who have the political and economic connections to "get things done." These latter groups are often not traditional landholders, and their major source of income is not from the land but from their business ventures. The rise of these groups has further weakened the power of the traditional village intermediary castes. Bailey has placed these groups in the middle class, since they transcend narrow caste and village interests and are involved in the outside world of politics and business.[5]

The effects of development upon the various nonfarming specialist castes—carpenters, smiths, barbers, laundrymen, and the like—have been mixed, depending upon the specific caste function. Some who were formerly marginal have gained greatly by the handicraft programs and have found a new economic basis for their activities—although these activities may be subsidized. Others have lost as their functions have been taken over by the impersonal forces of the market in urban areas or by an inflow of manufactured goods. Still others have abandoned their traditional client role within a single village, and have gained greatly in terms of both income and freedom by being able to take advantage of market forces in a group of villages or a nearby growing urban area to carry on their traditional occupations within a wider framework. Finally, some with useful skills have left the village to take jobs within factories, or have set up commercial manufacturing activities that utilize their skills. On balance there has probably been a net gain: the handicraft programs, the possibilities for employment outside the village, and the effects of higher incomes upon the demand for the craft skills and services have probably benefited rather than adversely affected the specialist castes as a whole, even if the effects of these and related developments might well be harmful to any one specialist caste.

Quite apart from the craft-specialist castes there are castes whose main function is trading and moneylending. Here again there may be gains or losses. On the whole these are the castes that are in business, and many have combined these activities with landholding. The effect of increased business activity in rural areas has been to increase the incomes of these

---

[5] On this see A. R. Beals, *Gopalpur,* especially pp. 58–63; and F. G. Bailey, *Politics and Social Change,* chaps i and ii.

caste groups; they have also probably gained as traders as a result of the regular but slow rise in farm prices during the Second Plan. However, they have lost somewhat by the introduction of new coöperative credit and marketing institutions, to the extent that they have been unable to use these to strengthen their own financial position; and, as pointed out earlier, various castes that were formerly pure landowning castes may have taken advantage of these new institutions to compete with the traditional moneylending castes. To the extent that these trading castes are numerically small, they also have lost politically by the introduction of democracy, and this political loss may have reduced their economic power.

The new position with respect to subordinate cultivating castes again varies from area to area, and generalizations are difficult. Land reform legislation has increased landownership by some tenants, or it has legally confirmed their tenancy rights to the land they cultivate and made possible eventual ownership. On the other hand, one of the frequently noted effects of the reforms is that landlords in many areas have forced tenants off the land in order to cultivate it themselves with laborers and to support their right of ownership under the reforms. Similarly, tenants have gained the least from the new financial institutions set up to aid peasants since they cannot offer land as security for credit. Within India, however, there is a mixed pattern of tenancy. Large landowners frequently lease land from smaller ones to increase their incomes, and smaller land operators lease it from large landowners. It is the smaller ones that I am talking about when I indicate these losses; the larger ones would probably increase their gains by their ability to rent more lands.[6] Furthermore, the effect upon tenants would vary with the farming background and experience of their caste.

The final group of castes to be discussed are the unclean or Harijan castes. Since almost half of the agricultural laborers were members of the scheduled castes in 1952–1954—or, reversing the data, approximately one-third of the members of the scheduled castes were agricultural laborers,[7] the conclusions that were reached with respect to agricultural laborers in the previous chapter in large part give a measure of the change in economic position of these castes. From the rough data available, it

[6] F. G. Bailey, *Politics and Social Change,* chap. ii, points out the gains to the tenant castes in Orissa; V. Dandekar, "A Review of the Land Reform Studies Sponsored by the Planning Commission," *Artha Vijnana,* Vol. IV, No. 4, December, 1962, has pointed out the negative effects; M. L. Dantwala, "Agrarian Structure," *Seminar,* No. 38 (October, 1962), pp. 30–32, stressed the mixed character of tenancy.

[7] See Table 2; see also D. Kumar's work cited in Appendix B pointing out the relationship between agricultural laborers and the former slave castes in the South.

seemed that in general these laborers probably remained roughly constant in their economic position. Kusum Nair concludes her book by stating that India, since 1947, "has not succeeded in changing in any essentials the power pattern, the deep economic disparities, nor the traditional hierarchical nature of inter-group relationships which govern the economic life of village society." [8] This is supported generally by the conclusion of such a specific study as H. Amir Ali's comparative analysis of a group of West Bengal villages in the period from 1933 to 1958.[9] This study showed only slight real improvement in the standards of living of all castes, including the scheduled castes. But there are two qualifications: first, the low castes in some of these villages have done better in maintaining their real income than the higher castes, and the income disparities between these caste groups have lessened; second, the proportion of low-caste children attending school has increased far more than that of other castes, in part because the base was so low in 1933.

In describing the position of the scheduled castes in the United Provinces, B. Cohn notes that many of the untouchable tenants were forced off the land when land reform was introduced; but he also makes the point that the widening of the economic interconnection of the village with urban areas led to a breakdown of traditional relationships and a job movement of untouchables to nearby urban areas and nonfarm jobs.[10] Mrs. Epstein, in describing the effects of the introduction of irrigation in Mysore, points out that in the village receiving water the traditional connection between higher and untouchable castes was reinforced as, after irrigation, the labor of the latter group became necessary; at the same time the relative economic disparity of the two groups became wider. In the adjacent "dry" village, the start of a nearby factory and the inability to get irrigation led to a rapid disintegration of the traditional patron-client relationship, and untouchables began taking nonfarming jobs. She also makes the point that, as new techniques or new output enter the village, the services needed, which are not included in the traditional patron-client relation, require some cash payment.[11] In some cases of such a breakdown of the traditional patron-client relationship, the untouchables and other members of client castes gain by being able to market their services

[8] K. Nair, *Blossoms in the Dust*, p. 196.

[9] H. Amir Ali, *Then and Now* (New York: Asia Publishing House, 1960), especially pp. 22–32, 57–67.

[10] B. S. Cohn, "The Changing Status of a Depressed Caste," *Village India*, ed. M. Marrott (Bombay: Asia Publishing House, 1961).

[11] T. S. Epstein, *Economic Development and Social Change in South India*, chaps 4, 7, and 8; also her article, "A Customary System of Reward and Improved Production Techniques," *The Economic Weekly*, Vol. XV, No. 11 (March 16, 1963).

in a rapidly growing factory or urban area; in other cases they lose even the mite of security that they had under the traditional system.

The most important gain made by the agricultural laborer, and probably the untouchable, is the elimination of *begar,* the requirement that the low-caste member supply free farm labor to the landowner when necessary. Prior to this abolition it was one of the normal client obligations, for which the laborer received at the end of the year a small share of the crop. As a result of the change the laborer is no longer obligated to perform unpaid work, and he must now be paid in kind or in cash for specific work that he does. In fact, this elimination of begar has been blurred somewhat; the voluntary labor (or *shramdhan*) parts of the community development program superficially resemble the old begar, especially in those villages where the farm laborers do the physical work on such projects as road building or well digging while the landowners supervise, but this is an exception. The effect of the elimination of begar means, "the force of hired labourers in Indian agriculture is now made up of free men. One could not say this a generation ago." [12]

The scheduled castes have also gained by the reservations of seats in schools, colleges, and professional and technical institutions and the reservation of jobs in government in proportion to their numbers in the general population. Although these reservations have not always been filled, the number of scheduled caste members being educated has risen more rapidly than the comparable number of clean caste members. A recent survey of primary school education all over India shows higher enrollment among Harijans than among the immediately higher castes. Furthermore, the number of scheduled caste members in government employment has also risen, although the number is still well below the proportions reserved. Finally, the Harijans have been getting preferential permits for building homes and other construction.[13] Progress has occurred in both these areas of education and employment, although it is halting and can be improved; and with this progress there has been some improvement both in the present economic position of the scheduled castes and, perhaps more important, in extended opportunities for education. In fact, the improvement has already been sufficient for the judiciary in several states to rule that reservation of seats in colleges and jobs in government in these states on the basis of caste alone is illegal; and

[12] D. and A. Thorner, *Land and Labour in India,* p. 8; see also S. C. Dube, *India's Changing Villages,* p. 135.

[13] See *Report of the Commissioner for Scheduled Castes and Scheduled Tribes, 1958–59* (New Delhi: GOI Press, 1959), Sec. VI on education and Sec. XIV on government employment; also "Progress of the Backward," *The Economic Weekly,* Vol. XV, No. 36 (September 7, 1963).

economic or class criteria must now be used in making these reservations. In Kerala, for example, the judge making the decision specifically cited the improved economic position of a low caste as a basis for eliminating caste reservation of seats in colleges.

In conclusion, I think it would be correct to say that the lowest castes have in general improved their economic position, with the one possible exception of tenant members. It would have been difficult for their economic position to have deteriorated, since the starting point was so low; but it may not have improved in some areas. The effects of the government policies and of economic development since Independence have been to improve somewhat both their present economic position and their potential for advancement, especially in relation to the Brahmins in the rural areas. At the same time the scheduled castes are still in general in the lowest economic position of any caste group, although they may be better off than previously.

SHIFTS IN ECONOMIC POWER: URBAN AREAS

I pointed out in Chapter 2 that the role of caste in urban areas is significantly different from that in rural areas. In urban areas the subcastes, or *jatis*, play a very minor role; broader caste groupings—simply the Brahmins, the other clean castes, and the untouchables—become more important. In the larger cities especially, communal differences reflecting origins and language differences are probably most important as divisive elements.

I think it is fairly clear that the Brahmins and other higher-caste groups, such as those included in the Bengali *bhadra-lok*, have lost in relative terms, and possibly even in absolute terms. As I pointed out, the Brahmins were the first to go to the cities. Without any business experience, and in fact looking down upon business, they would not enter business as a first choice; instead they entered government, and only as a second choice would they enter the clerical levels of business. In their position as government employees and white-collar workers, I think it fair to say, they have lost ground economically as government and white-collar groups have tended to lose. Many Brahmins, of course, entered professions that have expanded, and these may have gained; but that number is relatively small. The Brahmins, especially in the south, have suffered by the practice of job reservations and reservations for colleges and technological institutes. Brahmins have never been included among those castes defined as backward or scheduled to be given these privileges; at the same time since they were the caste most likely to go into government posts or into higher education when given the chance, they are the ones most adversely affected by the limitation. The Brahmins have also lost politically, which has helped reduce their economic power in fields where the state government decides, such as education and state employment. In Mysore, Brahmins might apply for only 20 per cent of the government posts; and only 30 per cent of the seats in medical and engineering colleges were allotted without reservation and thus open by competition to Brahmins.[14]

Of other communities within the urban areas, I think it probable that with few exceptions trading and business groups have gained substantially in economic power, whether they were of Brahmin or lower castes. These business groups have provided the capital and entrepreneurship in the most rapidly growing sector of the economy, factory industry. The extent of growth has varied among the groups, and so their relative position has shifted.

R. K. Hazari has indicated the caste breakdown of twenty of the leading business and industry groups within India.[15] Of these twenty groups, ten are from the merchant caste, two or three from the Brahmin, three are from powerful peasant castes, and four to five are casteless. In terms of community breakdown five are Marwari, four from Gujerat, three from Punjab, one from Maharashtra, one from Madras, one from Andhra, one from Bengal, two groups—including the largest one—are Parsee, and three groups are in whole or part English controlled. Hazari has presented an interesting body of data on the changing share that these twenty entrepreneurial groups have in the total gross and net capital stock of nonfinancial companies in both 1951/1952 and in 1958/1959. These estimates at current prices are given in Table 25. It shows the relatively higher rate of growth of the twenty leading groups—higher than other groups in terms of their control of physical assets. This is consistent with the noted shift in incomes in favor of high-income urban groups over the same period. However, within these twenty groups there have been shifts in position. The major change has been that the rate of growth of the purely European companies compared with the Indian companies is markedly slower. I think this reflects part government policy and part company policy in the sense that the European companies since Independence have naturally been somewhat conservative in expanding, and they have been willing to sell control of some of their enterprises to Indian groups. Among the Indian entrepreneurial groups the changes have been less noticeable. The Tata, Birla, and Dalmia-Jain groups, and the Martin

---

[14] M. N. Srinivas, two articles on "Backwardness," *The Statesman* (Calcutta) October 9 and 10, 1961.

[15] R. K. Hazari, *The Structure of the Corporate Private Sector,* chaps. 1-5; on the English companies, see M. Kidron, *Foreign Investments in India* (London: Oxford University Press, 1965).

TABLE 25
SHARE OF TWENTY LEADING GROUPS IN VALUE OF PHYSICAL ASSETS OF ALL PRIVATE SECTOR NON-FINANCIAL COMPANIES, 1951/1952 – 1958/1959
(millions of rupees)

	1951/1952	Change in Period	1958/1959
All companies			
(a) Net capital stock	12,530	10,960	23,490
(b) Gross capital stock [a]	17,360	14,680	32,040
Companies directly controlled by twenty leading groups			
(a) Net capital stock	2,581	3,787	6,368
(b) Gross capital stock	3,758	4,864	8,622
Proportion of leading groups control to total (per cent)			
(a) Net capital stock	20.6	34.6	27.2
(b) Gross capital stock	21.6	33.1	26.9

Note:
[a] Includes depreciation.
SOURCE:
R. K. Hazari, *The Structure of the Corporate Private Sector,* chaps. 1–5.

Burn group, which is only part Indian, are still the leaders and may have even increased their lead over the smaller groups. It is among the smaller groups that there have been some positional changes. However, Hazari's list does not go back to show the changes between Independence and 1951, which would show a relative decline in the Parsee community—excluding the Tatas—as a whole in industry; for several of the larger Parsee-owned enterprises were sold to other Indian groups shortly after Independence. There has probably been an improvement of the position of leading Hindu groups, especially Marwari, in large-scale business and industry. This is partly related to the more general shift in power since Independence from the very small European and Parsee communities to the larger Hindu communities including Sikh and Jain. These latter play now and played before a larger role in Congress party politics and financing than the non-Hindu groups. Another business group that is now rising in importance are the Sindhis. This is a Hindu refugee community from Pakistan that has a large business experience and background, and many of its members have been successful in Indian business. Although the position of these leading private business groups has improved somewhat within the private industrial sector, it is somewhat more questionable whether their position has improved statistically within the total economy. I previously noted that there has been a much greater growth in the government's share of the national wealth than in the private sector's share. This has been especially marked in the industrial sector. The Mahalanobis

Committee has also pointed out that the share of a Tata-controlled company, Associated Cement Companies, Ltd., in the output of cement has declined from 64 per cent in 1951 to 39 per cent in 1960; a Birla Company, Hind Cycles, accounted for 90 per cent of the bicycle output in 1951, but only 15 per cent in 1960.[16]

These business communities tend to be relatively small; and since many, such as the Marwari, play a national role in business, they are looked on with suspicion. Reaction to their expansion is mixed. Each state government prefers to give advantages only to entrepreneurs from its own state and may discriminate against "felt" outsiders. But at the same time they wish to develop new industries and to attract large private investors, who usually come only from one of these national business groups. Thus the policies of the states are very often at odds with themselves whether to encourage or to discriminate against these groups. In some states the net effect of discrimination is stronger and this discourages the larger out-of-state business groups. If at the same time local business groups in such a state are either financially weak or absent, the consequence may be a lagging rate of industrial growth if the slack is not taken up by government investment. In other states the preference for local industrialists may not lead to active discrimination against outsiders, and they may come in.

There is another interesting consequence of the discriminatory policies by the states against employing Brahmins in government and giving them seats in educational institutions. Brahmins have been entering business as entrepreneurs or high-level managers in larger numbers than ever before. This represents a reaction to the narrowing of traditional avenues of employment in government, but it reflects as well a demand on the part of the new business groups for the skills of administration and intellect that the Brahmins have always had. Brahmins have offered these talents throughout Indian history to the great chain of political rulers of varying castes and communities. Now they are willing either to use these skills in business themselves, or offer them to the new industrial and business power groups that may require them. At this time, however, their gains by this have unquestionably been exceeded by their losses in other sectors of the economy.

It is far more difficult to speak of other castes and communities. Communities engaged in industry have probably gained. The Sikhs, for example, have set up many small-scale industries and are working in factories; they provide many of India's taxi drivers and perform other mechanical services. They are willing to work with their hands, they like

[16] *Mahalanobis Committee,* pp. 36–37, 53–54.

to work with mechanical objects and seem to have an aptitude there; and in consequence many are successful. Some of the craft castes in West Bengal also have an aptitude for machinery, and they are employed widely in small-scale industries in that state. Other castes with no particular skills or with skills that are obsolete in handicrafts and industry have probably gained little, and may even have lost.

In urban areas the discussion of caste changes in economic power is consistent with the discussion both of urban class shifts in income and of rural caste shifts. But in urban areas, since caste lines have become blurred, it is much harder to make judgments on caste income changes than in rural areas. However, the general tendency in both urban and rural areas is a widening of the gains of development from a group that might accurately have been called minute to a somewhat wider group. This is similar to the shift in political power, and they are closely related. But this relationship from greater economic to political power is far from directly causal. Greater political power contributes to economic power probably as much as the reverse in contemporary India.

PART V

India's Future Economic and Political Trends

We are tied up in humdrum affairs, but a country with no vision gradually loses energy and perishes.

J. Nehru, Lok Sabha speech of August 22, 1963

"Well in *our* country," said Alice, still panting a little, "you'd generally get to somewhere else—if you run very fast for a long time, as we have been doing."

"A slow sort of country!" said the Queen. "Now, *here,* you see, it takes all the running *you* can do, to keep in the same place. If you want to get somewhere else, you must run at least twice as fast as that."

Lewis Carroll, *Alice Through the Looking Glass*

. . . No doubt the incipient movement in the direction of reorganizing national life on the basis of industry involved a breach with the customary methods of agriculture, which must in any case have caused a certain degree of dislocation . . . [But] the slow breaking up of the open field system . . . might quite conceivably have effected only such a gradual diminution in the number of the small farmers, as to make the absorption into industry of those displaced comparatively easy.

R. H. Tawney, *The Agrarian Problem in the Sixteenth Century*

11

Problems of the Economy

The main conclusion from our picture of the political and economic changes in India since Independence is that there has been a diffusion of political power and economic gains. This has meant greater power to village and small town elements in society and to the state governments. In the economic sphere the resultant diffusion of gain has come about probably at some sacrifice in the rate of economic growth. However, over the decade from 1951 to 1961, India achieved a rate of growth and a consequent sense of improved well-being that succeeded in keeping up with the economic desires of the interested population and, thereby, contributed to political stability.

At the start of the period a conscious economic development program was but an idea in the minds of several urban groups—the intellectuals, the administrators, and the industrialists. As a result of the two Five Year Plans and the numerous political and economic institutions set up throughout the country to carry out the development programs, the demands for development and the desire to achieve gains from the programs have become much more general. Furthermore, there is a much closer awareness of the relationship between economic development and government expenditures within an area so that not only have the aspirations risen, but so have the demands for government expenditures as a way to achieve them. Although the competition for expenditures has increased, the political limits on the government's ability to raise revenues by additional taxation are still narrow; this is especially so in any area where such taxation may affect the rural population. Assuming these things hold true, the implications of these conclusions for India's future are serious. It will become more difficult in the future to meet the economic aspirations of the Indian population, and the political risks of such failures will also become greater.

In this section of the book, I shall look specifically at the various sectors of the Indian economy in terms of their present situation and their problems. I shall examine various policies that might deal with those problems, not so much in detail, since that would require a far greater depth than would be warranted by this study or by the available data, but to present some insights into various alternatives from the viewpoint of an outside observer. Furthermore, my interest is not so much in the intrinsic worth of the policies, but rather in terms of their acceptability to the various political groups identified earlier, in the hope of determining areas of agreement and political possibility.

The most striking feature of the Indian economic picture since the Third Five Year Plan began is that the rate of economic growth, rather than accelerating under the impetus of the Plans and government expenditure, has in fact slowed down. Although the Third Five Year Plan envisaged a rate of growth of national income of about 5 per cent per year, in fact, national income has grown at only about 2–3 per cent per year, barely keeping up with the rate of population growth. The Plan envisaged a rise in agricultural output of 30 per cent; however, foodgrain output was stagnant at about 80 million tons, the level reached near the end of the Second Plan, during the three years from 1961/1962 to 1963/1964. It is estimated, however, that with the good monsoon in 1964/1965, output reached 88 million tons in that year; but with the very poor monsoon of 1965/1966 output is feared to have fallen to only 75–80 million tons, thus threatening local famines. In part, as a result, the wholesale price index of food articles has risen by about 40 per cent between March, 1961, and January, 1965, while the same indexes for other articles have shown much less change. Industrial output was planned to increase at the rate of 11 per cent per year; in fact it had risen at an average annual rate of about 7–8 per cent through 1963/1964 and then fell to about a 6 per cent rate during 1964/1965. There has unquestionably been little if any of the anticipated shift in the structure of employment, and open unemployment has increased during the first two years of the Plan. If the political future of India is in fact more closely bound up with its economic success or failure than even in the first decade, then the slackening of growth endangers the present political stability. At the same time, with the close relationship between economic policy and politics, the slackening growth also reflects political weaknesses that have become more and more apparent.

Certain rather obvious economic questions arise from this decelerated growth. Since agricultural output and employment bulk so large in the national totals of income and employment, why has the agricultural output shown this slow and erratic growth? Is it due only to the vagaries of the monsoons or harsh winters, or does it also reflect a misallocation of past investment in this sector combined with a waning of the incentives that stimulated output in the past decade? To what extent does the pattern of growth in agriculture influence the growth of industrial production? How can the level of savings within the economy be raised and the allocation of investment be improved? What effect has the shortage of foreign exchange and the government's allocation of the scarce supply had upon industrial output? What have been the consequences of the elaborate network of administrative controls and the previous unwillingness to use the price system as an allocation device upon industrial output? What can be done about the accelerating rate of population growth manifested in the 1961 Census, a rate significantly higher than expected by Indian planners? To what extent are the answers to these questions tied up with the changes in the Congress party over the decade?

Many of these questions cannot be answered with any precision. In spite of this they are questions that will have to be answered much more successfully than they have been during the early years of the Third Plan if India is to achieve the rates of growth and investment levels that are being intimated in preliminary discussions for the Fourth Plan—growth rates of 7 to 7½ per cent per year, and an investment level approximately double that planned for the Third Plan. This investment aims at achieving an ultimate target of a minimum per capita income of Rs 100 per month per family by 1975/1976, or an income sufficient to meet the "essential food, clothing, housing, education and health requirements of *every* person in India by that date." [1]

PROBLEMS OF AGRICULTURAL GROWTH

It is clear that if this goal of improved well-being for every person is to be achieved there must be an increase in the rate of growth in agricultural output and a more consistent growth pattern than in the past. Even with a drastic shift in the structure of the Indian economy it is most unlikely that, over the next decade and even longer, less than 60 per cent of the population will be employed in the agricultural sector, and 40 per cent of the national income will be derived from it. In absolute numbers, given the rising population, employment in agriculture will increase.

Any attempt to give a simple answer to the problem of increasing agricultural output would be naive and presumptuous.[2] However, it is possible to indicate certain potential policies for growth. First, although

[1] On progress in the Third Plan period, see GOI Planning Commission, *The Third Five-Year Plan;* GOI Planning Commission, *Third Plan Mid-term Appraisal* (Delhi: November, 1963), especially chaps. i and ii; on price movements, see "Behind Price Spiral," *The Economic Weekly* (Special No.; July, 1964), pp. 1293–1296; *The Economic Weekly,* XVI, No. 20 (May 16, 1964), pp. 831–832; also see GOI Ministry of Finance, *Economic Survey, 1964–65,* (New Delhi: February, 1965); K. N. Raj, *Indian Economic Growth* (New Delhi: Allied Publishers, 1965); and *The Economic Weekly,* XVII, Nos. 32–38 (August and September, 1965).

[2] My experience in India has been largely limited to the industrial, commercial, and urban sectors, so that I am hesitant in discussing agricultural policy and the role of incentives. Also in these few pages I am trying to summarize sharply; discussion of Indian agricultural problems could easily take a whole volume.

it is often stated that India's per acre yield of many crops is among the lowest in the world, this by itself cannot lead to a conclusion that Indian agricultural output can either be increased rapidly with a few technical tricks or that it is inefficient today. In fact, it is the opinion of some experts that, given the present availability of factors of production and their prices, Indian agricultural production is efficient; this contributes to making change so difficult. Furthermore, there is much evidence that Indian peasants are sensitive to price changes that affect their output: the rapid introduction of sugar cane production in North India in the 1930's; the shift to jute production in India and Pakistan both before and after Partition; the expansion of cotton output in the Punjab with irrigation prior to Partition, and the crop shifts in the use of acreage there; the influence of irrigation in Mysore State upon crop patterns and economic activity—all testify to the conclusion and support the hypothesis that Indian peasants are not very different from other peasants. They respond to new cost-price relationships, especially with regard to that portion of their output above subsistence.[3] This is not to say that they may include in their costs items not included in other countries; and the ignorance of nonvillagers with respect to these costs or to various social constraints may explain some of the peasants' unwillingness to adopt new techniques, or respond to suggestions from outsiders.

This raises the question obvious to an economist as to whether present incentives in India encourage both greater output of agricultural products and an increase in their sale in exchange for manufactured products? One of the alternative policies on incentives in the agricultural sector is essentially a policy to encourage those individual peasants with the resources and the skills to take advantage of new techniques, improved prices, and so on. During the decade of the first two Plans the increases in output in the agricultural sector tended to come largely from peasants with greater landholdings. However the slowing increases in output over the past three years raises some question whether the incentives and the new technological possibilities that were provided in the past decade are

[3] See on this, among others, R. Krishna, "Farm Supply Response in India-Pakistan," *The Economic Journal,* LXXIII, No. 291 (September, 1963), pp. 477–487; R. M. Stern, "The Price Responsiveness of Primary Producers," *The Review of Economics and Statistics,* LXIV, No. 2 (May, 1962), pp. 202–207; W. P. Falcon, "Farmer Response to Price in a Subsistence Economy: The Case of West Pakistan," *American Economic Review,* LIX, No. 3 (May, 1964), pp. 580–594; B. R. Chauhan, "Rise and Decline of a Cash Crop in an Indian Village," *Journal of Farm Economics,* XLII, No. 3 (August, 1960), pp. 663–666, describes the effect of increasing taxes upon tobacco production in a village; T. W. Schultz, *Transforming Traditional Agriculture* (New Haven: Yale University Press, 1964), pp. 44–48, 94–96 refers to W. D. Hopper's study of an Indian Village. D. Narain, *The Impact of Price Movements on Areas Under Selected Crops in India, 1900–1939* (London: Cambridge University Press, 1965).

still sufficient to encourage future growth. Essentially those incentives were some minor improvements in technology and inputs, higher investment in irrigation, and preferential access—to the members of the dominant castes anyway—to coöperative credit and marketing facilities and to some of the new inputs.

Higher farm prices have not been used as an incentive to raise general farm output. The relationship between higher farm prices and agricultural output in India is a complex one. When one speaks of higher farm prices as a policy instrument two meanings are involved—the conscious move to shift the money prices of farm outputs relative to manufactured inputs, or shifting prices of one type of farm output relative to other types of farm outputs.

Accepting this distinction among policies, within India there has not as yet been a conscious attempt to use change in farm prices to encourage farm output. At the same time, because of the poor farm outputs since 1961, farm prices have risen relative to other prices. Between March, 1961, and January, 1965, the wholesale price index of food articles rose by 40 per cent compared with 9 per cent for finished manufactured goods. The effects of changing farm prices upon output will vary depending upon the type of policy adopted. As pointed out earlier there is a good deal of evidence that Indian peasants are susceptible to changes in the prices of the various outputs that they do produce, and this has resulted in shifts in output among those products. These output shifts apparently also occur within a reasonably short period of time; Raj Krishna indicates that they occur in India as rapidly as in other countries. There is also evidence that changes in the relationship between the prices of specific inputs and the prices of the outputs they contribute to lead to changes in the use of those inputs—whether water, fertilizer, or whatever.

However, there is a good deal of question with respect to the effect of changes in relative prices of farm products and nonagricultural prices as a whole. Total farm output in India is still in large part dependent upon variations in the weather, especially the monsoon. A good monsoon will have a far more stimulating effect upon a farmer's output than any relative price change for the commodity, and more than the investment of that farmer. On the other side, a poor monsoon will have a harmful effect upon output, regardless of price stimulus, or the efforts any one peasant may take to prevent its harmful effects. Thus, changing price relationships as a whole will have relatively minor effects upon total farm ouput. At the same time the risky character of agriculture encourages speculative withholding and disgorging of farm output by peasants able to invest in stocks. In the short run, in fact, higher output prices can encourage such withholding, either in the hope of higher future prices or by improving the peasant's ability to hold off from selling. The new financial institutions have contributed also to the latter as I have noted earlier. There is some evidence that the agricultural and price experiences of the past few years support the above conclusions. Although the statistical evidence is unavailable, my impressions from a brief trip to India in 1965 is that the agricultural and rural sectors in general have gained relatively in income terms since 1961 as a result of the rise in farm prices compared with other prices. (Nothing is known of the distribution of the gains—it may be assumed that this follows the earlier pattern.) If this is true, it is believed to have contributed to the apparently greater political success of the Congress party in rural elections since 1962. It also stimulated the greater urban unrest that was obvious during 1965. There is also believed to have been greater individual investment by those peasants who did gain, and this was a factor in the greater farm output that has been estimated for 1964/1965. However, this favorable economic effect was undoubtedly overshadowed by the favorable effect of good weather in that year and of poor weather the following year. Furthermore, the increased financial resources of many peasants also made for a greater ability to withhold crops from the market for speculative purposes during these years.

For all these reasons I am skeptical of the use of price policy, in the above sense, to raise total farm output. In urban areas, such a rise in food prices would lead to demands for higher wages and higher industrial costs. That would have both serious political repercussions in the sensitive urban areas and possibly harmful consequences for India's competitive industrial position in international markets. The effects of such a price movement would also lead to a shift in resources away from the industrial to the agricultural sector, and thus in many ways contribute to slowing down the rate of industrial expansion in India. (These questions in India are very similar to those raised in the U.S.S.R. in the 1920's, which gave rise to the "scissors controversy" and the related discussion on the price question among Russian political leaders and economists.)

The previous discussion, however, does not say that improvement in the pricing mechanism would not be useful; one such use should be to provide greater stability of farm prices. The instability of farm prices probably discourages investment in agriculture to increase output, and it encourages greater speculation in inventory as a source of income. If the government, by a policy of stabilizing agricultural prices to the peasant, were able to reduce this source of risk, it would encourage peasants to greater investment and other efforts to increase output. Such a policy would also call for both widespread construction of public grain warehouses and a willingness by the government to buy for its stocks and sell from them to stabilize prices. It is argued by several Indian economists that the Indian government's use of U.S. Public Law 480 wheat supplies to reduce the wholesale price of wheat during the Second Plan period, in order to satisfy the urban consumer, had a discouraging effect upon subsequent wheat output in India. This is contrasted with the performance of rice output, whose price was not influenced by PL 480 supplies. If this is correct, it would have been preferable for the Indian government to have used the PL 480 wheat to build up wheat stocks when this was possible, and to have used those stocks to maintain stable, rather than falling, prices for wheat for both the urban consumer and the peasant. This would also be desirable policy for the future, once India's present food position begins to improve.

There is some question as to whether the present relationship between prices of inputs and prices of farm products is such as to encourage use of new inputs. For example, it is claimed by some economists that the price of the new fertilizers in India has in the past been so high relative to the price of farm products that peasants have been discouraged from greater use of them. With higher food prices in the last few years this ratio has become more favorable, and the demand for fertilizer has risen. Similarly the prices of other new inputs—water from new irrigation or new seeds—has often been so high as to discourage peasants from using them. For the peasant the risk involved in new inputs is very high; since his output is so much determined by factors completely out of his control such as the monsoon, the effect of a single new input, or even several, in any one year might be negligible, whereas the cost of that input to him would be high. For this reason, the price of new inputs, in their introductory stage anyway, should be such as to encourage their use, taking into account the risk element. This means that it may be desirable to subsidize their use initially, even though the prices of these inputs in the longer run should be such as to yield a desirable return on the investment made in producing them.

Another direct monetary incentive that would operate upon the returns of the peasant would be the use of higher land taxes to stimulate output and to encourage a greater flow of agricultural output away from subsistence purposes.[4] The direct effects of such higher taxes in stimulating total output depends on the peasant's desire to maintain his total pre-tax income—money plus subsistence in kind—after the new taxes are imposed. Since the larger commercial peasants, who would generally be most affected by the various proposals, have sufficient output from their plots to meet their subsistence requirements, the effect might simply be a reduced money income to them from the sale of their surplus; it might

[4] This has been urged by Ashok Mitra in "Tax Burden for Indian Agriculture," *Perspective,* II (June, 1961), pp. 1–27. The argument in the previous paragraph on the effect of PL 480 was made to me in July, 1965, by A. M. Khusro and K. N. Raj.

also discourage investment by these peasants. The smaller subsistence peasants would either not be affected since most tax proposals do not envision higher taxes for them, or they would be forced to sell a higher proportion of their output than at present and thus would have a reduced amount available for subsistence unless they could increase total output, which might prove difficult for them. The price proposals might be considered the carrot to encourage greater output, and the tax proposals would operate as the stick to force peasants to produce more to maintain their pre-tax incomes.

Since risk is such an important element in the considerations of the peasant, and so often the natural elements contributing the risk—weather, pests, and so on—are completely beyond his influence, consideration might be given to a crop insurance program that would cushion some of it. It might be possible to work out such a program, carefully considering its feasibility and cost, and then to try it on a preliminary basis in one or two districts over a period of years—both to measure its costs and its attractiveness over time and to determine the effects upon the peasant's willingness to try new techniques as a result of a reduction in a natural risk.

Related to the problem of incentives is the question of land reforms. Although the rights of the intermediaries have been eliminated and there has been some breakup of the largest estates as a result of land ceilings and other reforms, many states have not introduced or implemented land reforms that would confirm tenants into owners and improve the legal position of tenants. There is evidence that tenants, especially those on a share-cropping, short-term basis, are less willing to make the efforts and investment to raise their output than owners. Steps to implement existing land legislation and introduce new reforms could be especially rewarding in providing stimuli to increased output. The effect of land ceilings on output are potentially mixed. To the extent that they force absentee landlords to cultivate their land or turn it over to cultivators, they are desirable; however, they also limit the ability of the productive peasant to invest in additional land if he is near the legal size limit. There may also be economies of scale in peasant agriculture that are limited by the land ceilings.

In a related area of reform, perhaps of even greater importance than ceilings, are steps to encourage consolidation of plots. Many village holdings are badly fragmented into very small plots scattered throughout an area. Consolidation has been urged by the central government, but only a few states have been carrying out consolidation programs. Obviously such programs, which require rearrangement of plots and exchanges of land, are deeply involved in village and state politics and action has proceeded slowly.

Japan had great success in raising farm output from small plots of land until 1920. The Indian statistics on output in relation to plot size are highly dubious, but they do indicate that output per unit of land is higher on smaller plots than on larger ones. This obviously runs against the economies of scale argument, at least with present inputs and technologies, and it may reflect a higher labor input in the smaller plots. Furthermore, with the continued growth of population and employment on the land it is unlikely, even if ownership forms change, that the ratio of labor to land will be reduced or that the size of the unit of cultivation as distinguished from the unit of ownership will change greatly. India will continue to have a labor-intensive peasant agriculture with relatively small plots of land in the future, and the problem of Indian agriculture is primarily how to raise the productivity of these small plots of land.[5]

This raise the question of whether the possibilities opened by the past investment have been such as to permit major increases in output on existing plots of land. The Food and Agricultural Organization of the United Nations points out that the postwar irrigation schemes in the Ganges-Brahmaputra region of eastern India have aimed primarily at insuring against the failure of the monsoon during one season rather than at providing a year-round water supply. If the irrigation system did supply water the whole year, it would permit double cropping.

> For the survey area as a whole it should be possible approximately to double the annual harvested area within the present cultivated area. . . . The provision of an assured water supply would also provide favorable conditions for intensive use of fertilizer and higher-yielding varieties. From such evidence as is available, it appears likely that yields under farm conditions for the staple cereals, pulses and oil seeds could probably be approximately tripled under irrigation, provided it were accompanied by good farming practices. . . . This would make it possible to . . . provide a well-balanced diet for up to double the present population.[6]

There is some belief in India that these gains are possible not only in the above area but elsewhere with a heavy investment in irrigation, drainage, and flood-control facilities; but the rates of return on this investment are still unknown. At this point the question arises whether this additional investment would be preferable in agriculture or elsewhere. Certainly the returns in the recent past on total investment in agriculture have been

---

[5] One of the best analyses that I know of the problems of land reform and economic development in India is R. Krishna, "Some Aspects of Land Reform and Economic Development in India," *Land Tenure, Industrialization and Social Stability: Experience and Prospects in Asia,* ed. Walter Froelich (Milwaukee: Marquette University Press, 1961).

[6] FAO, *Possibilities of Increasing World Food Production* (Rome: 1963), pp. 61–63.

small in India. In fact the direct relation between agricultural investment and greater farm output is not precise; there seems to be agreement that a massive additional investment program without improved incentives on the lines discussed above would probably not achieve the desired results in output. A heavy additional investment in agriculture in the form of major irrigation and other works of long gestation would be a gamble in the present state of knowledge, and it might be wasteful of very scarce capital resources that could be used elsewhere. On the other hand, improved incentives and relatively small additional investments in research and so on, as I will indicate, might yield major output returns without massive physical investment. It is also my opinion that the present level of investment in agriculture, which is probably a minimum for political reasons alone, could be allocated in such a fashion as to yield higher returns. It could be done both by changing the type of irrigation projects, and by concentrating investment and other efforts in geographic areas that show high output potential rather than widely dispersing efforts and resources. On all these grounds, the desirability of greatly increased capital investment in agriculture is very questionable without a greater knowledge of probable results. This is even admitting that the investment in major irrigation works may prove less in real than in money terms insofar as it calls on idle labor resources and may not require much foreign exchange. The decision to embark on such projects should be made only after an educated estimate that found such additional investment in agriculture measured in real costs would yield a real rate of return commensurate with the yield on alternative investment in industry.

The question of heavy additional investment in agriculture should also be considered in the light of India's ability to continue to get large PL 480 aid in the form of food grains from the United States in the future, or its potential comparative advantage in industry as opposed to agriculture. If both of these are positive it would reduce the future gains to be derived from investment in agriculture and favor industrial investment.

Even without heavy investment in irrigation facilities, much can be done to improve the technology of Indian agriculture by providing new inputs. Indian agronomists themselves [7] have stressed the inadequacy of much of the research on new seeds; there has also been much criticism of the entire effort of the demonstration program, the poor communication

[7] See the report of the speech of A. D. Pandit, Vice President of the Indian Council of Agriculture, in *Eastern Economist*, August 16, 1963, p. 301; the newspaper report by H. Venkatasubbiah, "Spotlight on Agriculture," *Hindu Weekly Review*, August 12, 1963; the "Recommendations of the Agricultural Team," as reported in *Eastern Economist*, September 13, 1963.

among agricultural scientists, and the lack of coördination of the research efforts in the various states. Such research is essential before it is possible to apply technologies and inputs derived from experience outside India to the Indian environment. Without this research to adjust foreign experience to India, new inputs probably will not succeed. Their failure would not only be an immediate failure but could set back by years the whole process of introducing such inputs into a particular area. The Indian peasant simply can not afford to be burned twice by a risky investment in new seeds or inputs, even if they are given free. At the same time the potential yield from investment in agricultural research is high and it would not require the tying up of massive capital resources. It is obvious that the introduction of these new inputs calls for an improvement in the communication system between the research institute and the peasant. The present village-level worker is only rarely a good transmission belt of new research and technology. Experiments on a small district basis might begin to be conducted to improve the system of incentives to village-level workers for increasing output and to penalize poor work.[8]

Apart from investment in research the government is committed to a program of developing the fertilizer industry. Assuming that the investment yields a good return and the products are both adjusted to the quality of the land and the water supply, assuming that they are priced so their use is economic, development of this industry should have high priority. This investment would be industrial as well, and so it would assist in transforming the economy, stimulating agriculture, and increasing the flow of resources from the rural to the industrial sector. There may well be other manufactured inputs of a similar nature, such as improved but still simple plows, that would contribute to similar results. To encourage the wider use of new inputs that call for capital, expansion of the coöperative credit and purchase programs among the peasants would also be necessary.

Until now the discussion has been in terms of creating a climate of economic incentives and new technological possibilities that would encourage the small individual peasant to expand his output. Hopefully the effect of this program would be a revolution on the land—a revolution that develops peasants who derive income from the production of commodities, are motivated to produce additional output, and will reap the gains from the additional output in the form of higher incomes and larger land holdings. This would lead to the end of the jajmani system and its traditional social structure of rights and obligations not based on pro-

[8] See *The Economic Weekly*, 1964 Annual Number (February, 1964), pp. 343 ff.

duction. It would eliminate those people from the land who draw their income largely from social position rather than economic services.

I pointed out earlier that there is a group of influential and well-informed leaders and thinkers that disagrees with such an incentive program for individual peasant cultivators. This group argues that there will not be a large increase in farm output without a peaceful revolution in farm institutions that would eliminate private peasant production of agricultural products. This would hopefully unseat the landowning, moneylending, and trading groups that are considered to be dominant in agriculture, and, if combined with large-scale coöperative farming it would achieve economies of scale in agricultural output. The introduction of coöperative farming would be on a voluntary basis: it would not be the forced collectivization of the U.S.S.R. or China. Coöperative farming should be combined with, should be the capstone to, a broad program of land reforms to guarantee the peasant his land. There should be expansion of the service coöperatives that supply the peasant with credit and cheaper inputs; and as well there should be more help for marketing and processing his output. Last, there should be continued investment in improved farm technology and research.[9]

Although voluntary coöperative farming would appear to be a useful institution, its introduction in India would face very serious problems. Those coöperative farms that the government has set up have not been successful. There is a serious doubt in my mind whether the peasants could in fact successfully coöperate in such an effort in the light of the factionalism that pervades the village, and the administrative skill to run a large coöperative farm enterprise is rare. Apart from these political and administrative questions, there is evidence against the economy of scale argument used in favor of coöperative farming. All these points have been quite beside the fact that the strongest advocate of coöperative farming in the Congress party in the past was Prime Minister Nehru, that although the Congress party nationally had favored such a policy, the state governments had not implemented the recommendation, nor were they likely to.

In agricultural policy in the past there has been a stalemate between

---

[9] See D. R. Gadgil, "Planned Agricultural Development," *Opinion,* September 17, 1963, and his essays on agricultural policy in Gokhale Institute of Politics and Economics, *Planning and Economic Policy in India* (Bombay: Asia Publishing House); B. Singh, *Next Step in Village India* (Bombay: Asia Publishing House, 1961). For a critique see Raj Krishna, "Some Aspects of Land Reform and Economic Development in India," chap. v in W. Froelich (ed.), *Land Tenure, Industrialization and Social Stability* (Milwaukee, Wisc.: Marquette University Press, 1961), and his review of B. Singh's book in *Economic Development and Cultural Change,* XII, No. 1 (October, 1963), pp. 104-107.

the vocal and influential advocates of coöperative farming and stronger land reforms, and the landowners, party members, and state officials, who are not vocal but are influential in carrying out policy. The former are against incentives that would encourage the individual peasant; the latter have not been strong enough to prevent the adoption of past policy statements or to fight for an alternative national policy, but they have been strong enough to prevent the stated policies from being implemented. In effect the result is conflict and no general policy. Instead policy has been a mosaic of bits and pieces, such as to discourage investment and greater output and to lead to the present agricultural stagnation.

There are areas of agreement between both groups: that improvement is needed in the agricultural research program and in the communication program with respect to new technology, in expanding service coöperatives for credit, purchasing and marketing facilities, and in expanding fertilizer output. There would probably be little disagreement on a price policy for new inputs such as would encourage their use; both might agree on the desirability of adopting policies to reduce the price fluctuations in farm commodities. Both might well agree on certain types of land tenure reform—at least implementing the laws for the security of the tenant—so that the peasant has greater incentive to invest. These might appear to add up to a modest program, but it is more than has been accomplished up to now and would probably contribute to a more rapid rate of growth in farm output than has been possible heretofore. Both groups would probably agree on a massive additional investment program in agriculture—partly because of a genuine feeling that priority should be given to agriculture and partly for a desire for the political gains of such a program. The bureaucracy of the Irrigation Ministry, which would prepare and implement it, is also one of the oldest and most powerful in the Indian government. But to me such a program appears economically questionable.

It is most unlikely that either an extended coöperative farming program or greatly expanded land reforms will be achieved within the next decade. The existing structure of landholdings and ownership will very likely continue for reason of political fact. A real economic issue arises over whether the level of capital investment in agriculture should be greatly increased. On the basis of the low return achieved heretofore, I should say no, not unless proposed increases in investment promise a reasonable yield. At the same time I am certain the allocation of the present annual investment in agriculture can probably be improved significantly along the lines indicated in the FAO study; the possibilities of labor-intensive investments in agriculture to utilize underemployed labor should be explored further and implemented where feasible.

A separate question is one of increasing the flow of resources from agriculture to industry. To the extent that it is possible, I should advocate raising the land tax so that it is at least at the same real level that it was in the prewar period.[10] Such a step might also have good incentive consequences, but the political possibilities of this are quite limited at present. No political leader in India has sought to present a land tax program that would unite the small landholding and the urban elements that are not now affected by a land tax. Together these two groups include the largest portion of the voting population and they are already relatively heavily taxed; thus they have interests opposed to the larger peasants, who would have to pay the difference in higher direct taxes. A political coalition might be formed to raise land taxes in the same way as a coalition was formed to abolish the intermediaries. The difference of course is that now the larger peasants are politically strong; in the fight over abolition of the intermediaries many of the larger zamindars had taken the wrong side in the Independence struggle. In Russia the diversion of output from the agricultural to the industrial sector was carried out through the collective farm system, which served essentially as a forced taxation device; but it was only at a cost—which Russia found could be heavy—of lives and of failure to increase output. India is rightly unwilling to risk that at this time. The alternative lies in persuasion and in creating incentives that would induce the peasant to sell his product to the industrial sector; the higher profits of the industrial sector derived from these greater sales could then be tapped for industrial development. If the prices of industrial products were raised relative to agricultural products, it might have such a consequence; but its effect might be to reduce the quantity of manufactured goods purchased by the rural sector, so the total value of purchases even if the unit price rose would remain about equal to what it is now. Little is known of the elasticity of demand for industrial products by peasants. The effect of such a price policy upon farm output would also be depressing if, as I indicated, farm output is somewhat elastic to a price fall. Finally it could have serious political consequences since it would adversely effect the large farm population.

Consideration therefore should be given to the alternative of expanding output of those manufactured goods consumed by peasants. These could be either inputs for farm output, for example fertilizers and farm tools, or consumer goods, for example bicycles and radios, but they must be at prices that the peasants are able and willing to pay, which for consumer

[10] I. M. D. Little has presented an ingenious proposal for a progressive land tax that would both raise a substantial amount of revenue and support land ceiling reforms by creating monetary disincentives for the holdings of larger size plots. I. M. D. Little, "Tax Policy and the Third Plan," in Rosenstein-Rodan (ed.), *Pricing and Fiscal Policies* (London: George Allen & Unwin, Ltd., 1964).

goods may be much lower than present prices. On one level this would result in a shift in trade in favor of the peasants, and yet the increased quantities sold and the possible economies of scale from greater output would permit industrialists to maintain their present or even some higher level of total profits, even if unit prices of industrial goods were reduced in an effort to reach the rural market on a larger scale. It can be argued that in the Soviet Union, once the decision was made to embark upon a program of rapid expansion of capital-goods output at the expense of consumer goods, the incentives for voluntary increases in agricultural output and for the sale of these products to urban areas disappeared, and collectivization became unavoidable. In the Second Plan, it was hoped that a greater supply of consumer goods and inputs to the farm population would be achieved through the cottage industry program, which was a deliberate program to encourage labor-intensive output of consumer goods. In fact this program has not proved successful. Much of the output of the handicraft sector has been heavily subsidized; at the same time the output of the factory sector in competing goods has been deliberately limited to enable the handicraft sector to compete. Rather than serving either as a source of manufactured goods for the peasant or as a source of revenue and income for the nonagricultural sector, the subsidized expansion of the cottage industry sector has served as a mechanism for transferring government resources to the rural underemployed. The cottage industry program may have been justified as an unemployment program in the Indian context; it has had little other economic result.

A program for increasing the output of manufactured goods for the peasants could also be associated with suggestions for rural industrialization. It might be centered about the creation of rural townships to serve as market and production areas for the needs of a group of villages in an area; also, it might be used for further industrializing the small and medium-sized cities. The expansion of industries in such townships and smaller cities would reduce migration to the great urban centers and reduce the social strains associated with such migration.[11]

PROBLEMS OF INDUSTRIAL GROWTH

The previous discussion of incentives for agricultural output leads into the problems of industry and the expansion of industrial output. It is the industrial sector that has been growing rather rapidly; and it is the still faster expansion of the industrial sector that offers the greatest hope of drawing workers out of the low-income rural sector thus raising per

[11] John P. Lewis, *Quiet Crisis in India* (Washington, D.C.: The Brookings Institution, 1962).

capita incomes. Up till now that growth rate has not been sufficient to absorb the growing population in nonagricultural activities; and the growth has not in recent years proceeded at the expected rate.

What are the reasons for the disappointing performance in the industrial sector? One obvious possibility is that the stagnation in agricultural output has contributed to the slower growth in industry. K. N. Raj has pointed out that poor agricultural seasons and low outputs lead to higher prices of foodstuffs and of agricultural inputs, higher wages and higher costs, reduced profits and reduced incentives for expansion. The relation on the demand side is less direct, since the peasant sector today purchases relatively few manufactured goods from the industrial sector. The two main commodities purchased are cotton textiles and kerosene, and agriculture is more a potential than a present purchaser of such inputs as fertilizers.

The implications for industrial growth are that industry today expands on the basis of, first, urban consumer demand and then private and government investment demand. The limitations on the supply side are the real resources available for investment, including those that must be supplied from overseas. On the financial side, the domestic savings by the industrial sector, whether publicly or privately owned, and urban household savings provide the resources for industrial development; to the extent that their investment is shifted from industry to the rural sector or is misallocated to low-yield investments, the rate of industrial growth will be less. Since India does not produce much of the capital goods to expand its industrial capacity, it must increase its foreign exchange supply to import this equipment. She needs as well greater exchange to meet any additional foreign exchange requirements for imported raw materials and consumer goods arising from its higher rate of industrial growth and from its higher level of economic well-being. A continued increase in the supply of foodstuffs is required from internal farm production and imports sufficient to maintain the growing industry-employed and service-employed urban population.

It may be argued that the Indian government has not done enough to increase national saving by mopping up the demand for additional goods created by its own expenditures for development and defense. It is very difficult politically to hold Indian per capita consumption static: the growing population alone would increase the total consumption, but the rising expectations of the whole population and the fact that some social groups are improving their position and spending conspicuously add to the demands for greater individual consumption. Unquestionably, something more can be done by taxation, especially by raising the taxes on agriculture and the middle income groups and by more stringent enforcement of present direct taxes. Not much more can be done in this respect, given the present political system. This means that internal prices will rise; and they have done so, but at a rate that was until 1963 not unduly rapid. However, the annual rate of increase of the wholesale price index was about 10 per cent from March, 1963, to March, 1964, and this rose to about 15 per cent from March, 1964, to January, 1965.[12] These rises reflect both the constant farm outputs and the accelerated defense spending since 1963, which has risen to much higher levels than before. Such rising prices do create more favorable opportunities for Indian producers to sell in the domestic market and so discourage exports. Unquestionably too the pace of development and investment creates a demand for capital goods and other imports that exceeds, at least in the short run, India's own possibilities of increasing foreign exchange earnings, by either its internal financial policies or its production policies. (This is not peculiar to India; historically many countries required foreign loans during the early stages of their industrialization.) To a great extent India's increased foreign exchange requirements have been met by foreign aid and loan programs and by a series of drastic controls on imports and the use of foreign exchange to reduce the demand. But in the longer run the solution to this requires an expansion of India's own exports.

The expansion of India's exports is obviously a difficult problem; both the income and price elasticities of its three traditional exports—jute, tea, and cotton textiles—are probably low in overseas markets. This is apart from the fact that India faces increasing competition from newly producing countries and from synthetic fiber substitutes. It is difficult for a country to break into the world market with new industrial products. But India's own policies have been such as to discourage exports and in some degree to encourage imports. Given India's present exchange rates,[13] its slowly rising price level, and its fully protected market, there is little incentive for India's producers to produce for export. They can sell what they produce in the internal market, and there is no worry about competitive prices, quality, and so on. At the same time prices of imported goods are relatively cheap at the existing exchange rates, with a consequent demand for foreign exchange and a thriving black market for foreign exchange permits. Logically, there is a clear case for an outright revision of the present exchange rates, but the practicality of such a step is somewhat

---

[12] See H. Venkatasubbiah, "Runaway Prices," *Hindu Weekly Review,* June 1, 1964, p. 10; GOI, *Economic Survey, 1964–65,* pp. 29–32 and Appendix Tables 5:1 and 5:2.

[13] India's present exchange rate is about Rs 4.7 to $1.00 (or Re.1 = 21¢); W. Malenbaum, "Comparative Costs and Economic Development: The Experience of India," *American Economic Review,* LIV, No. 3 (May, 1964), pp. 390–399, suggests that a more realistic rate today is Rs 7 = $1.00 (or Re.1 = 14¢).

more dubious in terms of its effects upon other countries such as Pakistan or Ceylon producing competitive products, or in terms of the elasticity of foreign demand for India's exports or India's demand for imports. Similar results might be achieved by a program of export subsidy and import taxation, each of which would operate as a direct incentive or cost to Indian producers affected. Certainly, the foreign exchange component of investment should be valued at its present real rate rather than at its official rate in estimating the costs of, and the returns on, a particular project.

The argument that the weakness in India's export performance in recent years reflects policy weaknesses, more than simply world forces, is supported by a statistical analysis of India's exports. Study shows that its proportion of world trade in the traditional exports has declined while world exports have increased.[14] Apart from a program to adjust prices so that they reflect the real supply and demand situation within India, conscious steps are necessary to strengthen the Indian sales effort in foreign markets, to improve the quality of its output of exports, and to establish proper servicing procedures. These steps are especially important for the new engineering products that India hopes to export from the output of its new factories.

A source of foreign exchange that has not been tapped is India's great —and, through smuggling, increasing—gold hoards. India's gold price has been well above world prices. The rising price level for commodities and the low official interest rate encourages the holding of gold not only for traditional ornaments and jewelry, but also as an excellent hedge against inflation. If India were able to reduce the foreign exchange drain involved in smuggling and persuade the Indian population to turn some of its hoards over to the government, the gains to its foreign exchange supply would be most significant. An effort to introduce much more direct control over the holdings of gold and the manufacture of jewelry has been largely unsuccessful in part for political reasons; but even without this, its success would have been doubtful. Interest rates and foreign exchange rates set at levels closer to the real values of capital and foreign exchange to the economy would discourage both gold hoarding and smuggling, and this is a strong argument for such steps.

With respect to an increase in the supply for foodstuffs sufficient for the growing industrial labor force, hopefully some increase may be achieved from India's own output by the steps indicated earlier in this chapter. The continued supply of PL 480 grains and industrial raw materials, especially cotton, from U.S. surpluses offers a good prospect of maintaining food and textile supplies for the urban areas at a stable price.

---

[14] See B. Cohen, "A Comment on S. J. Patel's Analysis of Indian Exports," *The Indian Economic Journal,* Vol. X, No. 4.

This is provided they are used for such purposes in conjunction with Indian efforts to expand output and increase the flow of foodstuffs from agriculture to urban areas, or eventually increase manufactured exports to pay for food imports. Care must be taken in the release of these supplies to prevent a discouraging effect on output.

Although the internal financial resources for continued industrial expansion can come from industry and the urban sector, several questions arise with respect to the allocation of such investments. Those resources should be used in their most economic fashion so that returns on the investment are maximum, or at least very high. In the past this has not been the case.

The figures on the rate of return in government industrial investment are biased downward by the extended gestation period for the three major steel plants; but the rates of return on government industrial projects have certainly also been very low, when measured in money terms, unadjusted for either the real cost of foreign exchange or savings on imports. Professor Ramanadhan shows that the rate of return on nineteen government-owned industrial undertakings completed and in full operation averaged between 3.5 and 4.0 per cent from 1958/1959 to 1960/1961. Even with the three plants of Hindustan Steel in full operation by 1962/1963, it is doubtful that the rate of return has increased significantly. In a more recent study, the Reserve Bank of India shows that forty-five central and state government operating companies of an average age of ten years—excluding financial and investment companies, various started but non-operating companies, and Hindustan Steel—were earning a rate of return, measured by gross profits as a percentage on total capital employed, of only 5.5 per cent in money terms during the three years from 1959/1960 to 1961/1962. I am ignoring any external effects of such investment on the private sector. No estimate of this exists. Such low money rates of return, which are approximately one-half the same rate in private industrial undertakings, on government-owned gross assets of Rs 3.5 billion in 1961/1962 which are rising by over Rs 400 million per year, indicates either misallocation of scarce capital and foreign exchange as well as other resources, poor management, an uneconomic pricing policy, or all three.[15] This does not deny that some of the government firms, most notably Hindustan Machine Tools with its effective management, have been highly profitable.

This discussion also has certain implication for India's policy on the regional distribution of investments and allocation of industry. At present,

[15] See V. V. Ramanadham, *The Finances of Public Enterprises* (Bombay: Asia Publishing House, 1963), chap. i; *Reserve Bank of India Bulletin* (October, 1963), pp. 1267–1276.

as pointed out earlier, under pressures for equality of development in all areas, many policies have been adopted to ensure that every state is given equal access to modern or fashionable industries. Examples of such policies are numerous: the industrial location policies previously mentioned; the regional price equalization policy with respect to steel, which sets an equal price for steel throughout the country; and the effect of the policy on coal transport, which at one time made it possible to buy coal at a lower price in Bombay or Cochin than in Calcutta near the coal fields. The effect of these policies is to deprive potential low-cost industrial areas of their advantages, and encourage plants to locate in areas of higher real cost, to delay construction of plants that would be economical, to raise the cost of inputs to Indian producers of manufactured products as well as to peasants, and to make it more difficult for Indian producers to compete in world markets. Many of the states have encouraged the fashionable steel-using and industrial engineering industries; the need at this stage is a concentration of industrial efforts into fields, fashionable or otherwise, that yield maximum returns from the scarce resources and that, in turn, permit or encourage a maximum rate of industrial investment.

Both the continued outflow of resources from the nonagricultural to the agricultural sector, which reduces the supply of domestic capital for industry, and the apparent misallocation of resources in industry are major factors contributing to a rate of industrial growth below what has been expected. More has been hoped for by the planners, and more is essential to produce the necessary structural changes in the economy. Another of the contributory factors to this misallocation of scarce resources has been the reliance in the past on an extremely complex and time-consuming network of administrative controls over decisions on such matters as a firm's entry into industry and its ability to get foreign exchange or major raw materials, to raise capital, or to locate in one state or another. This has been combined with a fear of responsibility, indecision, and buck-passing, so that most bureaucratic decisions, even relatively small ones, have been made on high and have been subject there to political pressures and conflicts. The state governments are, if anything, even slower than the central government. In Mysore State a recent investigation concluded that it took approximately nineteen stages and nine months for a file on a matter to pass through the bureaucracy before a decision was reached; and also that almost all the petitions to one department need not have come up at all.[16] Over the past year the Union Minister for Finance and

[16] See on this, among many other recent articles, "Running the Industries," *The Hindu Weekly Review*, August 19, 1963; "Economic Controls," *Times of India*, August 16, 1963; and from a probably more sympathetic ideological position the

the Union Minister for Steel and Heavy Industries have begun to take steps to dismantle some of the controls and loosen others; but the past effect has been deeply hampering to industry, and there are enough remains of the past system to raise difficulties for entrepreneurs.

Apart from the direct economic effects on the returns from investment, the administrative effects of this network of controls have been most serious. Among India's scarcest resources are the administrative and entrepreneurial personnel experienced in, and capable of, dealing with large problems; many of these, in both the government and private sectors, are so tied up in handling bureaucratic problems of a relatively petty nature that they do not have the time or the responsibility to deal with major problems.[17]

Such a complex network of controls also contributes on the bureaucratic level to direct corruption, probably still not as great as is rumored about the higher levels; but even more important it contributes to indirect influence through job offers, marriages, and the like. By now this is sufficient to weaken popular faith in the integrity of the bureaucratic and political process and to encourage an "anything goes" attitude in a growing part of both the urban intellectuals and the rural population. At the same time the complexity of controls encourages the very monopoly and inequality that the government officially opposes and supposedly discourages by other controls. Without proper contacts, without a willingness to spend a great deal of time in New Delhi or the state capitols and to tie up relatively large sums of capital and time in travelling to Delhi, it is very difficult for a would-be entrepreneur to enter industry with any size plant. The largest firms or groups—already on the spot with Delhi establishments, with contacts on a political and administrative level, with relatively ready access to quick finance, and with many alternative uses for their resources—operate at a great advantage; and as I have previously shown they have grown relative to other firms or groups. It is doubtful, too, if the controls, especially on price and distribution, do in fact achieve their stated goals. From all reports there is a well-established "gray market" in many of the controlled commodities; and the effect of the controls is not to prevent misallocation, but to give intermediaries profits in these com-

article by P. K. Chaudhuri, "What is Wrong with Our Administration," *The Economic Weekly*, Vol. XV, No. 32 (August 10, 1963); and "Defreezing Steel," *The Economic Weekly*, Vol. XV, No. 33 (August 17, 1963).

[17] I am speaking of industry in this section, but many of the same problems of confused and overlapping administration, of undue dispersion of apparent responsibility, of lack of power, and so on, also plague the agricultural program. See, among many, Venkatasubbiah, "Runaway Prices," and the remarks of the Minister for Agriculture in *Statesman*, February 18, 1963.

modities which are either illegal outright or nearly so, and obviously to encourage evasion and corruption.[18]

How can these controls be dismantled when the Indian economy is so short of many resources, and has such large demands on those scarce resources? I strongly believe that the past unwillingness to use the price mechanism as an allocative device contributes directly to the network of controls. One of the price system's major purposes is specifically to perform the allocation function. It may be argued that—under certain conditions, such as large external economies, or in the event of a massive maldistribution of wealth and income—it may function either incorrectly or contrary to certain desirable social goals and thus give rise to significant differences between private and social returns.[19] The present, almost complete abandonment of the price system as an allocative device leads directly to great pressure on administrative resources and serious misallocation of resources requiring still further controls. The question is not one of abolishing all controls, it is one of using controls in conjunction with fiscal measures to supplement the price system rather than abandoning it; this should be done not on ideological grounds, but on administrative grounds. It is significant that the U.S.S.R. is discussing wider use of the price system as a stimulant to output; it appears its greater use is not inconsistent with socialism.

The Indian distrust of the price mechanism stems from many grounds: among the officials of the government it arises from experience with controls going back to the war years; among political leaders and the public it arises from a general feeling that prices for necessary items of consumption or development should be kept low, in the belief that costs of development are thereby lowered and equality encouraged; and in both groups it is related to a Fabian bias that prices are somehow capitalistic and inconsistent with socialism and that efficiency is unimportant. Because of the flagrant black marketing of business groups during the war and the relatively low social status of businessmen vis-à-vis Brahmin government officials, the distrust of prices reflects a distrust of businessmen. An additional contributory factor is the belief that planning and development is a scientific, mathematical process[20] that uses physical balance and input-output techniques in which prices are supposedly unimportant and do not influence the outcome of the plan. Also, the long existence of these con-

---

[18] On this see chap. 7. On the growing problem of corruption see the speeches by the Home Minister, Mr. Nanda, referring to the increasing corruption, and its infiltration into higher levels than heretofore, *Statesman Overseas Weekly*, November 9, 1963, p. 13.

[19] The economic literature on this is abundant.

[20] This may also reflect Prime Minister Nehru's own science background.

trols has built up vested interests by now of both officials who administer the controls, whose jobs depend upon them, and of business groups that gain by the possibilities of black or gray markets in the controlled products.

If many of these direct controls can be eliminated by the far greater use of prices to perform the allocation, and by such steps as simplifying procedure so that the acquisition of one permit for a new enterprise ensures that the other permits are forthcoming—let us say the foreign exchange permit is the key—the consequence will be not only to reduce red tape and delays, but to improve efficiency of performance on the part of the economy and possibly even to reduce monopoly by permitting freer entry into industry. A first step was recently taken to decontrol distribution of steel; recently, too, the prices of sixteen commodities were decontrolled. There are other steps that would be useful: the setting of a realistic rate structure for railroads would contribute to easing the demand for space for the movement of such key items as coal, and would thus also assist in dealing with the coal problem. The cement, cotton textile, sugar, and coal industries, among others, are also heavily controlled with respect to various aspects of their activities; consideration and study might also be given with respect to their decontrol, in whole or in part, in the near future. To the extent that prices are set, they should be fixed in relation to other prices and they should be geared to a development goal—say higher relative prices for coal—in order to attract capital to some industries and discourage its flow to others.[21] Finally the official money rate of interest on capital used by the government in calculating the cost of its own investment, as well as serving as a basis of the commercial bank rate of interest charged to industrial lenders, has been kept at an artificially low level, especially when compared with the rising price level. Together with the low foreign exchange rate it has had the effect of stimulating an un-

---

[21]Alan Carlin has examined the economics of the Indian railroad rate structure in his unpublished Ph.D. thesis, *An Evaluation of U.S. Government Aid to India* (M.I.T., June, 1964), chap. 3. I have not had a chance to see the World Bank sponsored study on transportation. William Johnson, whose study of the steel industry will be published shortly, has concluded as follows:

> Although the prices of certain categories of steel have been decontrolled, there is some question in my mind as to how effective these reforms will be. First, only the prices of relatively abundant sections will have been freed. Prices of scarce forms, which are most in need of revision, continue to be fixed by the government. Second, prices not established by the government are now set by a committee upon which the government has sometimes, and the Railway Ministry (as the representative of consumers) has often, exercised veto power. According to an informant, this committee has thus far been unable to function as it is supposed to function. To be sure, the principle of decontrol has been established. Yet there was steel price decontrol in the early 1950's, only to have controls reestablished in 1955 after steel scarcities recurred. Present decontrol may be short lived.

economic demand for capital and encouraging the use of capital-intensive technology. It may have contributed also to a reluctance of private individuals to put their savings in government bonds or into banks; much saving has been put into gold. The upward adjustment of the interest rate in the organized capital markets, and the use of a realistic shadow rate in estimating the cost of capital invested in planned projects [22] would be most desirable to reduce the misallocation of scarce resources. Continuing on this series of decontrol steps permits the market to assume a greater role in the allocative process while limiting controls to key areas where they can be administered most effectively. Plotting the necessary steps requires both detailed analyses of particular markets—of the kind carried out by the Raj Committee before the decontrol of steel—and a decision to adopt a general priee policy, rather than a hodge-podge of particular, and often inconsistent, controls. It also calls for increased fiscal efforts to mop up excess purchasing power created by government development and defense expenditures.

The greater supply and the improved allocation of foreign and domestic resources in industrial investment should also contribute to better use of present Indian industrial capacity, much of which operates today, as in the engineering field, one-third below full capacity. The approach to fuller use of capacity would also contribute to reducing costs as economies of scale became operative. Second, where present capacity is being operated at or near 100 per cent, and the economic prospects are favorable, the government should seek to expand industrial investment and output at a higher rate than heretofore, either in its own investment program or by incentives to private businessmen. This may call for continued expansion of the economic overhead sectors—power and transport—in association with industrial programs, when power and transport are major bottlenecks. It calls for continued expansion of the capital-goods industries—such as steel chemicals, and machine tools, when they have economic justification apart from being merely import saving. It also calls for the development of consumer goods, especially those for the rural population, and farm input industries produced by mechanical power. The stress has been on expansion of the capital goods section in the past, but it is rural demands that have been neglected with unfortunate consequences; to attract greater output from the peasants, and to transfer it to urban areas, the peasants must be offered goods that they want and are able to buy. The savings on foreign exchange and on domestic resources, including the administra-

[22] W. Malenbaum, *American Economic Review*, Vol. LIV, No. 3 (May, 1964), estimates 15 per cent as a realistic shadow rate of return on capital; Carlin estimates a 10 per cent rate. Both are well above the rates used in the past.

tive, from not greatly enlarging the agricultural investment program, might be used for industrial expansions on these three fronts. Although I am not greatly optimistic about India's chances of expanding new exports in a single Plan period, I think many of the Indian planners are too pessimistic over a ten- to twenty-year period. The experience of Europe and Japan since the war, although not fully comparable to that of India today, does provide evidence that countries can enter or re-enter world trade successfully, and that the markets in developed and underdeveloped countries will respond even though under pressure for protection to competitive products from domestic producers.[23] The main criterion for planned allocation of resources among industries should be their expected rates of return, which can be computed only after making some corrections in present prices for social cost and benefit since the price system is so distorted. This probably would support import-substituting industries, but the start or expansion of the latter per se should not be the major criterion.

Policies should be directed toward mopping up purchasing power with higher taxes, increasing resources devoted to industry, and allocating those resources on primarily economic criteria via the price system. This combination would have the effect of accelerating industrial growth. It would operate both by reducing the gestation period for industrial projects between application for licenses and completion and by maximizing the return from those projects; thus plans could be made for a greater plowback of resources into industry in the following period.

POPULATION PROBLEMS[24]

There is little very new that I can say on the population problem. If India were able to reduce its present rate of population growth, it would gain economically by being able to shift resources from meeting consump-

[23] It was less than ten years ago that books and articles were being written proving the existence of a long-run dollar shortage because of the continuing superiority of the United States in productivity. This situation has changed obviously and drastically. On export policy and exchange rates see the series of articles by J. Bhagwati in various issues of *The Economic Weekly* in 1962, presenting the logic of a change in rates.

[24] The outstanding books on the subject available to the American audience are K. Davis, *The Population of India and Pakistan* (Princeton: Princeton University Press, 1951); and A. Coale and E. Hoover, *Population Growth and Economic Development in Low Income Countries* (Princeton: Princeton University Press, 1958). S. Enke, in chapter 20 of his book *Economics for Development* (New Jersey: Prentice-Hall, Inc., 1963) presents a scheme of cash incentives to encourage smaller families. He also refers to his earlier articles on this subject. (This is looked at with some skepticism by population specialists.) S. N. Agarwala, *Attitudes Toward Family Planning in India* (Bombay: Asia Publishing House, 1962), has reviewed a variety of studies on this subject in India. See also *Statesman Overseas Weekly*, November 9, 1963, p. 10.

tion needs to raising investment. The gains from development would be more rapid on a per capita basis. Until now the Indian government has supported population control, but it is doubtful that it has received a high priority. Crude evidence for the lagging program exists in the fact that although the Third Plan budgeted Rs 270–500 million for this program, only about Rs 50 million had been spent in the first three years of the Plan. Research is going on, in India and elsewhere, on developing inexpensive, simple, safe, and socially acceptable birth control devices. Recently, new intra-uterine devices have been developed, and these are believed to have a good possibility of success. The key problems are first to communicate the results of this research to the Indian population and to sell them on the new devices, and second to create a set of incentives to convince the Indian population of the desirability of limiting the birth rate more directly and in a much shorter time than in the developed countries.

Much more research is necessary on the relation between social attitudes and birth rates in India. Finally, it would be worth conducting some experiments, at first on a small district basis, with various types of birth control devices to determine the most acceptable one in different areas. These tests might be combined with advertising programs and with promotional schemes, such as that presented by S. Enke which provides for cash incentives for smaller families or penalties for larger families, such as a poll tax that includes children. If the government by its policies succeeded in developing a program that reduced the rate of population growth to the rate that existed at the start of the Plan period, it would have more beneficial effects for the well-being of the population than much of its direct investment in the output productive sectors. The population control program has received low priority in terms of administrative skills, power, and prestige, and therefore it has lagged in an operational sense. The development of such a program and its implementation is a task that can best be done by the government and could not be carried out effectively by any private organization—unlike some of the more directly output-producing efforts involved in economic development.

DEFENSE IN INDIA'S ECONOMY

Politically I have treated India as largely self-contained. In fact, India has played a large role in international affairs since Independence, but it was a role that did not impinge significantly upon its economy during the first two Plans. India's foreign policy was directed by its goal of non-alignment with any side in the cold war, and within this goal to provide a policy lead for the other newly independent countries and a moral lead for the great powers: such policies did not require a large military establishment on India's part. In the narrower military field, policy stressed defense against Pakistan, especially in relation to the Kashmir problem. The latter effort did not require major military expenditures,[25] and India was able to carry out its development plans with relatively minor diversion of resources to military purposes and without involving the military in economic planning for development.

With the attack from China in October, 1962, the situation changed drastically. India was defeated in the border skirmishes, and in effect lost territory she had claimed, even though the Chinese army withdrew from some territory it had overrun. Without entering into the broad consequences of this for its foreign policy, the economic consequences were reflected in both immediately increased expenditures and in new allocations from the budget for 1963/1964. Before the Chinese attack, the annual current defense expenditure ran at about Rs 3.0 billion, which was approximately 30 per cent of the total central government budgeted current expenditure and about 2 per cent of net national output. In 1963/1964 the level of current expenditure was budgeted at about Rs 7.0 billion, over 40 per cent of the total of such central government budgeted expenditure and more than 4 per cent of the net national output. Budgeted capital expenditures rose from Rs 0.5 billion to Rs 1.6 billion. Although actual expenditures for defense on both current and capital account ran somewhat below the budgeted amounts for 1963/1964, there was a more than doubling of the previous year's totals.[26]

India is now modernizing and expanding its military forces substantially, recruiting eight additional divisions for its army, and possibly doubling its total armed manpower from 550,000 to one million men. It is also planning to set up six new ordnance factories, and considerable progress has been made on one.[27] This will involve a substantial drain on India's resources, including the key limiting resource of foreign exchange, during the entire Third Plan and thereafter. It has been estimated that over the entire Third Plan period the investment in defense will be on the order of Rs 4.5 billion, and meanwhile current expenditures for defense will

---

[25] This was partly because Pakistan itself was influenced by the alliances of which it was a member, even though it received large military aid from those alliances.

[26] On this see *The Economic Weekly,* Vol. XV, No. 10 (March 9, 1963); and Vol. XV, No. 11 (March 16, 1963); GOI, Central Statistical Organization, *Estimates of National Income, 1948–49 to 1961–62* (New Delhi: 1963); the 1963/1964 budget with supporting papers; *Eastern Economist,* XLII, No. 13 (March 27, 1964), p. 711.

[27] For the data below see the National Council of Applied Economic Research, *The Economic Implications of the Present Emergency* (New Delhi: May, 1963), especially pp. 7–11, 43–45. See also *The Hindu Weekly Review,* October 28, 1963, p. 5; and *Eastern Economist,* March 27, 1964, p. 712.

rise by approximately Rs 12 billion. This is an amount equal to about 15 per cent of the original total Third Plan expenditures; such a large effort will require, perhaps most importantly, foreign exchange beyond that originally planned for economic development—an underestimate to begin with —on the order of Rs 3.0 billion, or $600 million. This $600 million is approximately 12–15 per cent beyond the planned foreign aid requirement of about $5.0 billion during the Third Plan. Apart from the demand for additional foreign exchange, the expansion of the armed forces to the extent indicated will call for approximately 15,000 engineers, craftsmen, doctors, nurses, and so on, who already are in short supply for the civilian economy. It will also result in a drain on certain key raw materials, such as steel, copper, and wool, which are also in short supply and must be imported.

This obviously provides both a challenge and an opportunity. The sudden increase, it is hoped, will level off and mesh into the economy in such a way as to stimulate industrial output, especially in engineering products, and to contribute to transportation requirements. However, these additional demands may give rise to a chaotic condition that seriously upsets existing industry. Their effect so far may have been more upsetting than stimulating: the rate of growth of industrial output has increased only slightly at best, and the increased defense expenditures have contributed to an unusually rapid rate of price increase since 1963. It has been observed that certain of the shortages and pressures with respect to coal and transport have relaxed, but this may reflect as much a declining demand for these key industrial inputs in the face of the lagging fulfillment of the Third Plan, as an increase in their supply or improved allocation. Until recently little progress had been made in relating economic defense planning to general economic planning. The *Economic Weekly* reported in 1963:

The emergency has snapped whatever tenuous link there was between the Defense Ministry and the Planning Commission. . . . [Whenever] local and foreign experts have investigated idle industrial capacity in a particular sector, they have been startled by what they have found out. Yet, the Ministry of Defense Production . . . has yet to make a thorough survey of what can be done to meet its needs by existing workshops. . . . Manufacturers continue to visit Delhi in futile attempts to speed up decisions on defense production. Others wait patiently for frozen stocks to be cleared and for the payment of long overdue bills.

Even in early 1964 the *Eastern Economist* still stated:

There has . . . been no adequate effort for full utilization of installed capacity even in industries, directly or indirectly connected with defense preparedness.

On the contrary, there are many industries of military significance which are working at partial capacity and there are few industries, indeed, which have been called upon to do an extra shift or even work up to the limit of normal shifts.[28]

To some extent the challenge was met in the financial area by the 1963/1964 budget, which raised taxes substantially and introduced new techniques of compulsory savings. The defense expenditure level for the 1964/1965 budget remained roughly constant. Furthermore, steps were taken to mobilize gold hoards. Although the details may be argued, the 1963/1964 budget took courage, as did the gold control, and it is to be regretted that the steps taken have been seriously weakened as the sense of emergency diminished and pressure from the state leaders against directly taxing the peasants under compulsory savings made itself felt. The main production area in which the stimulus of the Chinese threat is strong is in the expansion of government-produced ordnance output, the value of which is estimated to have doubled in one year. But in other respects there has been little stimulus to the economy, and there has probably been a net increase in confusion.

The solution to this aspect of the defense problem is only partly economic. India has very able economists, and its administrative structure would permit planning to tie in the economics of defense with that of development. The internal political situation, especially conflicts among the ministries, has up to now prevented dealing with the defense problem as part of the general economic problem, and it has mired all efforts at achieving any beneficial consequences from the emergency.

It is most likely that the change in the character of the confrontation with Pakistan over Kashmir from one of a cold war to one of actual fighting in August, 1965, will increase the pressure of defense upon the Indian economy. It may also force the necessary steps to better integrate the defense effort with the general development effort.

INDICATIVE OR DETAILED PLANNING IN INDIA?

Past planning has performed a major service for India's economic development, by setting goals that have stimulated public enthusiasm for

[28] *The Economic Weekly,* XV, Nos. 28–30 (Special No.; July, 1963), p. 1101; and *The Economic Weekly,* April 20, 1963, p. 661. I also recommend other issues of this magazine during the period from October, 1962, to July, 1963, especially the articles by R. Thapar on the Delhi atmosphere. See the press release of the Federation of Indian Chambers of Commerce and Industry, dated February 1, 1963, for a business viewpoint as of that date. *Eastern Economist,* March 27, 1964, p. 713.

development, creating an awareness of development problems, and encouraging efforts to contribute to their solution. More important, the plans have committed the government to major increases in expenditure and investment; and the economic overhead investments of the government, together with thc promise of a guaranteed demand for the goods used in government investments, has provided private business enterprises with a reasonably firm basis for their expansion and investment programs. Thus the favorable effects of planning have arisen from the indicative character of the Plans and their psychological consequences as much as from the actual government investments themselves.

My questions with respect to planning in India do not arise from these effects, but rather from the detailed character of the Plans. It has obviously been possible to plan for the government sector and government investments. Even here, however, for a variety of reasons the plans have been approximate, with a marked tendency to underestimate both the costs and gestation periods of the projects. Detailed plans—including very specific output and investment targets for industry and strict type-of-output prescriptions for the private sector—have had much less significance. In both agriculture and industry, very little is known about how the sectors function and what the relation is between them. The statistics that form the basis of the detailed plans are highly approximate, and communication from the Planning Commission at the Center to the state governments, the local entrepreneurial groups, and even the various operating ministries, who also are at the Center, is delayed and erratic. This makes planning in detail a very difficult task with the results frequently not reflecting the plans

Much of the reasoning behind the detailed plans is ideological. It rests on the grounds that such planning is part of "democratic socialism," and that it is "scientific"; both of these have been accepted by the leaders of the Congress party and by the administrative bureaucracy. The goals of improved well-being for the population, equality, and democracy are all praiseworthy; these, it is felt, are best achieved through socialist democratic planning. The rationale of detailed planning is also supported by a distrust of the businessman and the price system on the part of the intellectuals, the administrators, and frequently the higher castes.

In fact this ideology has been in conflict with the pragmatic requirements of planning and development. The government simply has not been able to take on many of the tasks considered ideologically desirable; but in order to prevent this gap between desire and practice from being misused by business groups, the government has introduced its complex network of administrative controls. Private businessmen involved in the web of controls have learned to work with them, and they have done so either in order to start and run the enterprises they want or to strengthen their own position vis-à-vis potential competitors. I think the government is correct in recognizing the difficulty of operating a socialist society within India, but the present system gives India the hoped-for benefits of neither socialism nor capitalism. Rather, the Indian government is caught in the middle between its ideological position and its requirements of growth. The effect of this conflict on the agricultural field has already been noted; in the industrial and economic national field it has contributed to overadministration and misallocation of resources from an economic criterion, and the goals of equality may have been made more difficult to achieve than otherwise.

A compromise is possible. It needs to be based on indicative planning techniques and broad goals set for the economy and the major sectors. Within this broad framework, plans for government investments can be precise; and it would even be desirable to broaden the present scope of planning to include analysis and policies that deal with the impact of defense preparations on the economy and to revise it to lay greater stress on population planning. But plans for the private sectors in agriculture and industry need not and should not be detailed. Rather the targets should be in terms of general output levels placed within the general context of a price system that reflects the real costs of resources to the economy and provides incentives for both the desired sectoral allocation of resources and growth for the private sector. The government might also set up general institutions—credit, technical, etc.—that encourage growth in desired directions. Given this general framework, the goals of equality and reduced concentration of wealth can be approached by an effective tax system directed toward those goals and by a vigorous antimonopoly program. Rather than an elaborate network of controls that prevent entry and discourage growth, the government should seek to achieve its equalitarian goals—after greater output has been achieved—to reap and redistribute some of the gains.[29]

The economic ideas with which the Congress party took power that provided the basis for the successful policies and plans in the past have worn out their usefulness in all but the broadest sense. The excess capacity of 1950–1960 in many fields has been utilized; the key problem is to stimulate growth in economically efficient areas. There is a need for another set of ideas, though perhaps they will be expressed in the words of

[29] One of the main arguments for control of entry into industry is the argument that, with economies of scale, only "optimum size plants" should be allowed. This discourages entry by smaller producers, and only makes possible entry by larger ones. It would be generally preferable to allow anyone who wishes to enter, take his risks, and if profitable, expand to a lower cost size.

the original ideology since those words may be politically acceptable. Ideas are needed that will assist the present political leaders in dealing operationally with India's present economic problems. The fact that nationalization of banks is proposed seriously as a main step toward approaching socialism and dealing with the problems of economic growth is an indication of the gap between the past ideas and the present situation. It speaks well for the Congress party leadership, including Nehru who was then prime minister, that the proposal was not accepted at the 1964 Congress party meetings. A far more fruitful suggestion was put forward by one of India's foremost economists, a convinced socialist, in a private conversation when he said that the problem is to convince the political leader that the price system is ideologically progressive. Without a new set of ideas in almost all policy fields and a willingness to drop old dogmas and experiment, the Fourth Plan will become largely meaningless. To plan for doubling the level of investment and raising national income by 7½ per cent, when it has increased by only 2 per cent for recent years, without detailed analysis of both specific factors that contributed to the 2 per cent growth and of the alternative policies to overcome these factors, is an invitation to continued failure. Such lack of realism can lead to the discrediting of the Plans, not only as operational programs, but as indicative guides and psychological stimuli.

12

The Political Trends and Their Implications for Economic Policy

POLITICAL TRENDS

To examine the policy alternatives within India I shall set forth the political probabilities, in terms of both internal developments and foreign pressures. Much more than one chapter could be devoted to this subject, even when crystal-gazing is restricted to a decade.

First, I think the Congress party will remain in power during this decade, under any conditions short of military defeat or economic collapse. This in effect means that the Congress party will have a reasonably effective leadership, that it will win the 1967 election, and that there will be no serious split within the party before the end of the decade.

On Nehru's death the political leaders of the Congress party, which means the major state leaders, united to select Lal Bahadur Shastri as his successor. Upon Shastri's death in January, 1966, the leaders selected Mrs. Indira Gandhi, Nehru's daughter, as Shastri's successor. Although there have been these rapid changes in the head of government, I expect the unity of the Congress party to be maintained during the decade of the 1960's. There is far more to keep the Congress together than to cause a split, for as long as it stays united it can most likely retain national power. However, with the frequent changes, power has been relatively weak at the Center. Shastri dealt courageously with the issue of Pakistan and consolidated his position by his leadership during the fighting with Pakistan; but he had only begun to exert his stronger power when he suddenly and unexpectedly died. Mrs. Gandhi is still an unknown quantity as a Prime Minister, and it would be foolhardy to attempt any predictions of her behavior. However it is unquestionably true that much of her effort will necessarily be devoted to the consolidation of her position at least until the next national elections in early 1967.

The fact that the state and party leaders chose Nehru's two successors increases the strength of the party and the states vis-à-vis the central government. Shastri was and most probably Mrs. Gandhi will be more influenced by the party than Nehru was; and those economic and social groups in village and small town society, as well as the new urban industrial and management groups that are now dominant in the party and the state governments, will increase their power in the central government. The connection of those urban intellectual groups who were still attracted to the party by Nehru's presence will probably be diminished. However, Mrs. Gandhi's leadership, with her background and prestige, should serve as somewhat of a substitute to that of her father in retaining the support of the urban intellectual middle class groups for the Congress party, at least through the 1967 elections and probably beyond.

I do not expect other democratic political parties to play much more of a direct role than they have heretofore, but they may play a greater role as a source of ideas and influence upon the Congress, as urban intellectual groups become less influential. The outright socialist parties recently united, but some of their former leaders have left them and joined the Congress and the government, and thus they may provide new ideas in the Congress. The Swatantra party, philosophically right wing and favoring the abolition of planning and fewer controls, proved somewhat stronger than expected in the last elections, and it has strong financial support. But unless it can transcend its identification with the Brahmins, princes, and large industrialists, I doubt that it will become a real national challenge to the Congress. The threat of its influence, however, may encourage the Congress party to reduce controls.

With respect to broad policy, although socialism will still be strongly supported by Mrs. Gandhi, it is likely that the party leaders will be less committed than Nehru to that ideology, and they may be more willing to sacrifice the ideology of equality for more rapid growth as a policy goal. However much the ideology might be sacrificed, the political pressures from the states for allocation of expenditures for political purposes might become greater so that in the end there would be little change. Nehru, as a national figure, was able to resist some of these regional pressures; his successors may have more difficulty in doing so, especially since they will have served only short periods as Prime Minister. There may also be some loss of the vision and the receptivity to ideas that Nehru had, but here again this may be somewhat compensated for by health and administrative capacity.

Until now I have looked at the potential political changes in terms of their effects on economic policy. There are some more independent political variables that may have changed with the deaths of Nehru and Shastri, at least in the early stages of Mrs. Gandhi's position; one is that the North will have lost power somewhat vis-à-vis other areas of India. The leading state and party figures today are those from the South, the East, and the West. The Congress parties in Uttar Pradesh, Mrs. Gandhi's state, and the Punjab are much weaker. The effect of this will be to shift political power away from the North and toward the South; one would expect this to be associated with economic decisions. With the South's greater apparent political power in India, however, the southern demands for independence of greater autonomy should be less, and this should contribute to a more unified India.

So far as intercaste relations are concerned, Nehru's position against casteism in social and political life and in favor of removing the discrimination against untouchables was very prominent. Mrs. Gandhi's position as her father's successor is likely to be equally strong, although that of the party leaders is likely to be less morally committed. Although the probability of further social reform in the near future is unlikely, the Congress party will remain committed to the present reforms. However, the more important problems in this field are likely to arise from possible consequences of continued economic stagnation upon these intercaste relations. If economic growth is not accelerated, intercaste tensions will become stronger. The lower castes will be expecting more, but they will find opportunities for jobs closed; the upper castes will be under greater pressure to maintain their positions in the face of population pressure and higher prices. The total effect would be to sharpen the role of caste and communal conflicts in political affairs.

An internal problem that has once again become stronger is the Hindu-Moslem communal problem. The Hindu-Moslem riots of 1964 were on a much larger scale than expected, and they reflect the increasing tension between India and Pakistan. On this basis, they may also reflect some of the increased pressures for jobs and status associated with the economic stagnation and difficulties of recent years. Fortunately the 1965 fighting between India and Pakistan did not result in communal rioting; and the united response of the Hindu and Moslem population to the conflict may contribute to reducing future tensions. Mrs. Gandhi is as strongly committed to the secularism and amicable Hindu-Moslem relations as her predecessors and can be expected to take strong steps against internal conflict. The economic effects of such communal riots are much less than the political effects, but they are important, both immediately in terms of loss of life and property, and in the longer run by discouraging minority communities from putting roots down in an affected area.

This communal problem is obviously associated with relations between India and Pakistan in Kashmir and in East Bengal. The amicable solution of the Kashmir problem would do more to place the Hindu-Moslem relations on a firmer footing in India than anything else. The release of Sheikh Abdullah from prison in Kashmir and his conversations with Prime Minister Nehru and President Ayub Khan indicated some groping toward a settlement. If one had been achieved, Nehru's prestige and leadership might have helped to make a settlement palatable to India; his death made it more difficult since Shastri lacked Nehru's prestige and power during most of his leadership. The reconfinement of Sheikh Abdullah and the outbreak again of large-scale fighting over Kashmir showed clearly the sharp deterioration of relations since Nehru's death. Shastri had, just prior to his death, succeeded in negotiating the Tashkent Agreement with Ayub Khan. This was an initial step toward putting relations on a better footing after the fighting. Shastri's death will make it difficult to take advantage of this initial step to proceed further in the near future, since Mrs. Gandhi's position as leader of the government will be weak, at least until the 1967 elections.

The worsening of Indo-Pakistani relations in conjunction with a greater Chinese threat have strengthened the demands for a larger Indian army with more modern equipment and for increased defense production. To the extent that the problems between India and Pakistan are resolved, these defense demands on the economy would be diminished, or the forces could be used along the Chinese border if need be. I expect that with less personal prestige and the changing character of the world, Nehru's successors will devote less attention in the realm of far-flung foreign relations and world affairs. This should, however, contribute to greater attention to India's relations with its immediate neighbors, including not only the strengthening of amicable ties but possible joint action in the event of attack.

The major threat to democracy during this period would be another series of successful foreign attacks in the Himalayan-Kashmir area.[1] I think that their success would seriously threaten Indian democracy and make a replacement by a military dictatorship, probably associated with right-wing nationalist and communal support, more likely. Such an event would certainly result in greater stress on defense production than previously. It would probably be immediately accompanied by some greater puritanism with respect to corruption—not necessarily less corruption. Possibly there would be some weakening of the politician's role as opposed to the soldier's and administrator's. I see little reason to think that it would lead to a higher long-run rate of economic growth than heretofore, mostly because in a right-wing communal government the traditional rural groups would be even stronger than in the Congress party.

A possible eventuality from any military defeats that would be even

[1] I think major attacks by the Chinese are unlikely, assuming that India makes no effort to recover the Aksai Chin area. At the same time one can assume continued Chinese efforts to win over and increase its influence in the border states along the Himalayas. Such efforts may even be stimulated by the increase of tensions with Pakistan over Kashmir.

worse than the previous one, would be the disintegration of India into various small linguistic groupings and states. A forced breakaway of one state as a result of a military defeat could lead other states to feel that the weak military protection they receive from the Center is not worth the cost of unity. This of course would be disastrous for India's future and would make the entire discussion of this book irrelevant. I feel however that it is a most unlikely contingency in the period under discussion and can safely be neglected.

I think there is very little probability of an internal Communist takeover in India during the decade either by peaceful or violent means. The Communist party itself lost prestige as a result of the Chinese attacks, and it is split between pro-Chinese and pro-Russian wings. The democratic parties are far stronger and will maintain their greater strength. At the same time the army would most probably crush any internal revolution, if it did not move to block an anticipated one.

In summary, it is probable that the Congress party will maintain its position as the governing party in the next decade; but the conservative and rural groups in the Congress party will grow somewhat stronger. Externally the demands of defense will become greater in the decade in response to the greater impact of China and Pakistan. At the same time the expectations for economic improvement have become stronger; and the stagnation of the last several years raises serious questions for future economic policy.

FUTURE ECONOMIC POLICIES

On the basis of the previous analysis I should expect that the economic policies of the Indian government will be somewhat less influenced by ideological factors. Without Nehru's power and influence they are likely to be somewhat more conservative, reflecting the shifts in power within the party. This may be reflected in a decline in the position of the Planning Commission, which was Nehru's creation and one of his major interests. As a result the operating ministries should become even stronger than they are now. This will strengthen the administrators at the expense of the economists, and will make it more difficult to look at the economy as a whole. The administrators may be more pragmatic than the economists, but they are less likely to look beyond the interests of their ministries, and they may have a stronger vested interest in the maintenance of administrative controls as an allocative method. However, there may be greater pragmatism in administering controls since the administrators are "operators." Among the new political leaders influential figures may be more opportunistic and more operational in their approach than Nehru was;

and they should be more willing to try new devices and methods, provided they are convinced that the use of these devices will not lose votes. But whether these adjustments will add up to a more consistent national program than heretofore is of course an open question at this time.

Within the agricultural sector, on the basis of past experience it is unlikely that increased direct taxes on agriculture will be used as a major source of additional revenue; further major land reforms in India or the extension of coöperative farming are also unlikely. Any deliberate "wager on the rich" is also ruled out for political reasons. However, there are certain minimal policies in the agricultural field that are compatible with either of the two viewpoints on proper agricultural policy that were outlined in Chapter 11. In the straight agricultural sector such minimal policies would include: more effective research on new inputs and outputs, and steps to improve the communication of such research developments to the peasants; continued expansion of the network of service coöperatives; and expansion of the small labor-intensive improvements in the villages such as minor irrigation works, construction of irrigation channels, contour bunding, and others that can yield high returns for small real investments. With respect to such labor-intensive improvements, the U.S. PL 480 grain shipments could be usable for partial wage payments in kind. The continued expansion of manufacturing industries producing such farm inputs as fertilizers and simple tools and the sale of these to the peasants at prices that encourage their use would also sitmulate greater farm output and increase the flow of farm products from the rural to the urban sector. Similarly, the expansion of industries producing manufactured consumer goods for the peasant market would provide incentives for higher farm output.

In fields of farm policy more sensitive to politics, steps might be taken to accelerate the present program to consolidate village holdings into unified plots. Finally, the government, by accelerating its program of constructing warehouses and storage facilities for holding grain at harvest time, by offering a stable minimum price for food products, and by buying and selling its stocks of grain—including PL 480 stocks—to maintain stable prices, can discourage speculative hoarding of foodstuffs which contributes to price fluctuations. Any such price stabilization program may face the active opposition of rural moneylenders, traders, and peasants who profit by sharp price fluctuations; they are politically strong in state and local branches of the Congress party. Great political courage may be necessary to limit agricultural investment by the government to those geographic regions and that type of investment where returns are largest. The political pressures from both the states and the bureaucracy for widely dispersed investments, irrespective of rates of return, are great and continuous; and they may result in misallocation of resources without compensating gains in other respects. These problems are not ideological. They can be dealt with through the preparation of a workable program and the political mobilization of sufficient support to carry out the individual steps within it.[2] Willingness to experiment with new ideas is needed, for example, in crop insurance, which might reduce some of the risk in new investment and techniques and thus contribute to higher output.

The process of increasing agricultural output in India is a difficult one. There is a long history of well-known and economic techniques, an understandable suspicion of the effectiveness of new products and techniques in the face of the great natural risks to production, a social system that emphasizes obligatory relationships above greater output, a fear of social changes that may accompany the steps to increase output, a suspicion of both increasing wealth among peasants and of new government advisors, and distrust by the government of the price system. All these contribute to the difficulty of raising agricultural output and reduce the potential return from major additional investments in the field. This raises serious questions about the economic efficacy, in the short run at least, of massive rural investments, and it underlines the desirability of institutional changes in agriculture including continued research and new incentives with the attitudes they encourage; and changes in these may be a slow process. If India can be insured of the continued supply of western grain during this process of transition, the urgency of increased farm output is lessened. However, this can be only if the aid does not decrease the incentives for change. But the problem of raising more resources from agriculture still exists; and with higher taxation or collective farms ruled out, the only alternative is a flow of industrial products that will persuade peasants to exchange their surpluses.

It is in the industrial field that more rapid growth may be expected and is required. A major barrier today is the network of controls that inhibit rather than encourage growth. The government is sufficiently pragmatic to have begun to recognize the need for change, and individual ministers at various times have urged reducing controls and have taken steps to do so. The problem is to convince a wider group of political leaders and administrators of the importance of the problem and the specific desirability of both reducing the controls and using the price system more effectively on a general basis. Political influence of industrial groups that favored fewer controls on industry is nowhere near as potent as the political influence of dominant agricultural groups in the agricultural sector.

---

[2] They provide opportunities for what Hirschman calls "reform-mongering." See C. E. Lindblom, "Has India an Economic Future," *Foreign Affairs,* XLIV, No. 2 (January, 1966), pp. 239–252.

There are vested interests in both the bureaucracy and the business world that gain from controls and gray markets; at the same time there is a large and influential group that finds controls a nuisance, even though they may have gained from them, and would welcome easing of the control system. The problem is also partly ideological: it is necessary to convince the political leaders that red tape and socialism are not necessarily synonymous, and that using prices and fiscal methods effectively as allocative devices is consistent with both socialism and greater growth. Although India's tax rates are superficially high, stress should be laid on the fact that an improved tax and fiscal system—in law, but especially in administration—is a more effective method of achieving goals of equality and higher output than controls that inhibit output and in fact reduce equality. Thus although ideology is a barrier to the easing of controls, if the arguments for the latter are well supported and presented on an administrative rather than an ideological basis, I believe a program of relaxation can be prepared that would be politically acceptable.

Again there is recognition of the need for rural industrialization, for expansion of exports, and so on. The problem is to convince the political leadership of the desirability of encouraging this more by specific actions and incentives than by words, and to examine and propose steps to be taken on a pragmatic basis to deal with the difficulties involved. The political obstacles represented by groups that might be against steps to improve policies in these areas are not the most important ones; rather the obstacles are in the administrative and political decision-making mechanisms. Difficulties may arise in the top bureaucracy, with its experience and commitment to direct controls and its belief that heavy industry and import-saving per se are most important. The desired change again is to approach investment choices on an alternative basis—one based on estimates of expected return, without a pre-existing bias. India probably does have a comparative advantage in many heavy industries, but they have no intrinsic virtue in and of themselves.

Very broad changes in policy are unlikely to occur quickly and it should not be expected that they will be accompanied by verbal ideological changes. What is politically feasible, and what might occur with the support or leadership of the new Prime Minister, is the development of specific and workable policies and their implementation in the less spectacular areas of policy—farm incentives, new fiscal steps, relief from the network of controls to industry, and direct incentives to exports. The new central leadership may eventually be able to exert the powers of the Center somewhat more strongly than recently. Strength is needed to resist more effectively the pressures of political groups operating through the states for widespread dispersion of industrial and agricultural investment;

likewise it is needed to curb the pressure of powerful bureaucratic elements for increased allocation of funds without examining economic returns, for example, to railways and irrigation, and for continuation of controls that hamper rather than stimulate growth. In the short run, without Nehru's prestige and strength at the Center, the states may be demanding more openly special privileges. Apart from these curbs, greater pressure should be exerted to ensure that programs for defense expansion are meshed with programs for development and that other programs—for population control, export promotion, tax collection, curbing of monopoly, and so on—are strenuously implemented. Present performance in all of these lags far behind hope, and exhortation frequently substitutes for action.

Such political leadership combined with several good monsoons can lead to a breakout from the relative stagnation in the total output of the economy over the past few years. There would also be a faster rate of increase of industrial output, a real change in the structure of the economy, and progress in dealing with the major short-term problem of unemployment. The pressures for defense induced by new open conflict with Pakistan may contribute to the political steps necessary for economic growth by the present government. As the changes do take effect the resources going into further industry will expand; possibly even more important, the political power of the agricultural groups in society will be relatively weakened, and the political basis for policies to encourage a more rapid rate of industrialization will be strengthened. But it is obvious that this relatively modest range of politically feasible alternatives will not permit spectacular changes in total output or in the structure of the economy in the short run. It is doubtful whether the supposed policies for such spectacular changes are compatible with democracy in India where the bulk of the voting population is rural; it is also doubtful whether spectacular shifts in the economy are possible even with a dictatorship and its heavy political costs.

Without steps that would lead to a breakout from the relative economic stagnation of recent years, the political outlook for the 1970's would be much gloomier. With continued economic stagnation one could expect continued deteriorating conditions on the land as population pressure becomes greater and industrial employment opportunities show little improvement. There would be no increases, and in fact there would be declines in real wages and peasant incomes while the glaring contrasts between wealth and poverty would continue. Although education and literacy would increase, so would "educated unemployment" and disillusion with charges of corruption and inactivity. Communal and intercaste tensions would increase as the struggle for the limited openings sharpened.

The consequences for political stability and future democracy in India could be disastrous. The probable victory of the Congress party at the next election provides approximately another decade for new ideas to be developed and for policies to be adopted and implemented to prevent this stagnation and to create once again both a sense of growth and continued gains from growth. This is probably the most important economic and political requirement—a necessary though not sufficient condition—for the maintenance of democratic political stability in India beyond 1970. In my opinion, there is a fair chance that this requirement can be met with U.S. assistance.

PART VI

Conclusions: United States Policy

The problem of foreign aid is insoluble if it is considered as a self-sufficient technical enterprise of a primarily economic nature. It is soluble only if it is considered an integral part of the political policies of the giving country—which must be devised in view of the political conditions, and for its effect upon the political situation in the receiving country.

H. Morgenthau, *A Political Theory of Foreign Aid*

13

United States Policy

The political picture in India indicates another five to ten years of relative stability. The threats to this in the near future are external aggression or, internally, an unexpected splitting of the Congress or a sharp deterioration of the economic situation. In the longer run, the continuation of the present economic stagnation could seriously threaten India's present governmental structure and stability. The probability of a five to ten year period of stability, however, gives time for the development and application of policies both to strengthen India's ability to meet external threats and to jar the economy out of its present rut. For both of these purposes, U.S. policy may make some difference.

In this chapter I am attempting neither a detailed presentation of a policy nor an elaborate critique of present policies. Both would require an extent of treatment far beyond my present purposes or knowledge.[1] I shall instead be dealing with broad policy strategies derived from the previous analysis.

I believe that from the point of view of the United States the present democratic system of government in India is desirable in terms of realistic alternatives and the intrinsic values of self-government. The success of democratic government in India is closely dependent on more rapid economic development and a more rapid change in the structure of the economy than has been achieved in the recent past. The United States and other western countries can help India to achieve these economic changes and their associated political results.[2] Apart from this relationship, the United States can also assist in dealing with any external threats to the present Indian political system. But admitting all this, the U.S. role is limited: it may help protect India against external attack; it may provide foreign exchange and technical expertise to supplement India's own capital resources and knowledge; and it may provide some advice on

[1] Alan Carlin has just completed such an economic analysis of American foreign aid in India; John P. Lewis in *Quiet Crisis in India* has devoted the last third of his book to examining problems of the present aid program.

[2] Although I feel that this is the case in India, I am not at all certain that a similar relationship between economic change and democracy or political stability holds in other countries. To determine this it is necessary to look at the individual countries independently and in similar detail as I have done with India.

policy. However, its direct influence on the internal, short-run politics, the power structure, and the political forces that are largely independent of economic and international policy is small. It is worth stressing this, because too much should not be expected from U.S. policy.

What is the security interest of the United States in India? I think the minimum is to retain a friendly, even if unaligned, India. Charles Wolf, Jr. has attempted to suggest measures of the value of a given country to the United States under certain circumstances and suggested, as one measure, the alternative cost of achieving the same results in the event of the loss of that country.[3] Without going into the theoretical basis of this, India would be rated very low in a central war situation with the U.S.S.R., but it would be rated high in dealing with Communist expansion in South Asia, whether the expansion is through external attack or internal subversion. This assumes that South and Southeast Asia are in themselves of value to the United States. A hostile, pro-Communist India —or one that had broken apart into linguistic states—would probably make the cost of maintaining the noncommunism of the block of South Asian countries from India to the Pacific prohibitive; and it would certainly make it more difficult to maintain friendly or neutral governments in Pakistan and Afghanistan. It would also raise the costs for defense of the off-shore island nations of Southeast Asia, and the defense costs of Australia and New Zealand would go to much higher levels than at present. Its psychological effects in such areas as far removed from India as Japan and South Korea would be profound. And it would have unfortunate psychological consequences in Africa and in those countries of Europe with long-standing connections in India or strong Communist movements.

Apart from these increased security costs, there would also be comparatively minor costs to the United States and its allies arising from probable changes in India's foreign trade pattern and from the loss of American investments in India. In 1960 these investments were on the order of $200 million to $250 million for the United States, and over $800 million for the United Kingdom; in 1961/1962 the American exports to India were about $450 million and imports from India approximately $225 million.[4] To the extent that trade between the countries would be reduced for political reasons, there might be somewhat higher costs to the United States for substituting imports from other countries for those from India,

---

[3] C. Wolf, Jr., "Some Aspects of the 'Value' of Less Developed Countries to the United States," *World Politics,* XV, No. 4 (July, 1963), pp. 623–635.

[4] Reserve Bank of India, *India's Balance of Payments, 1948/49–1961/62* (Bombay: 1963), Table 11, p. 61 and Table 33, pp. 136–137.

although India would probably still wish to export to the United States. For those American exports not associated with aid, there would be somewhat reduced gains to the United States from alternative markets. The likelihood of repayment of loans made since Independence would obviously be much less. As a positive effect, the economic aid from the United States to any future pro-Communist India would undoubtedly be reduced; this would represent a resource saving to the United States.

If we assume, at the other extreme, not a pro-Communist, but a communalist India governed by a military dictatorship allied with Hindu communal groups, the costs to the United States would be less than in the previous alternative, but still higher than at present. It is possible that the position of India with respect to China would be stronger; and India would in fact present a greater barrier if it did not provoke military conflict with China by attempting to recover lost ground. However, relations with Pakistan would probably be worse even than the present situation. The present Kashmir issue with its constant border skirmishes might become exacerbated, and the U.S. effort to maintain peace in the area would become more difficult. The cost of maintaining effective treaty relations with Pakistan would be higher, since relations between Pakistan and China would probably be closer. The costs of defense to the Indian economy would probably rise substantially and divert increasing resources from development, which possibly could have destabilizing long-run internal consequences.

For both of these possibilities, it is in the interest of the United States to help a united India maintain its democratic government against the alternatives of a Communist or a right-wing communal dictatorship. As I have stated earlier, I think in India's case failure to achieve a sense of economic improvement—not necessarily a great statistical improvement or an improvement in comparison with other countries—would contribute to such dictatorships. The Congress party long ago made economic development one of its main goals, and the people are aware of this. The successes of the past decade have only whetted the appetite for this goal, since the past decade has primarily created preconditions for further growth. At the same time the increased literacy with wider education, the continued inflow of population into the cities, the increasing pressure on the land, and the demand for employment have probably increased the desire for economic improvement. It is significant that the recent sample surveys on the Indian state of mind before and after the Chinese fighting indicated that the Indian population's main hopes and fears on an individual level centered on the standard of living and opportunities for children, and that hopes on a national level revolved around hopes of economic improvement. These obtained even though the fear of Chinese aggression increased greatly after the fighting.[5] Failure to create a sense of improvement in economic conditions would contribute to political changes that from our point of view would be politically undesirable.

Even if there were no Communist governments in either China or Russia or any security interest in India for the United States, I still believe that the United States, on moral grounds, should assist the Indian government in its attempt, necessarily difficult and slow, to raise the Indian population from its poverty. The fact that the Indian government is democratic provides further grounds for assisting it in its efforts. Such a government permits its people a greater degree of freedom and mobility than any dictatorial alternative. As a democracy itself the United States should be sympathetic to such a democracy and willing to support it to some extent when it offers both evidence of past success and hope for the future. In the case of India, the security arguments of such an assistance policy support the moral argument.

The policies to achieve such goals are not just limited to an economic aid program in India. They include internal U.S. policies on foreign trade and overseas investment; they include military aid or support for policies; and, finally, they include a range of policies to influence India's internal decision-making apparatus. However, potential U.S. influence in internal decisions is peripheral and probably indirect. Any U.S. policies are necessarily limited in their consequences. They may serve as a life net preventing the collapse of the Indian effort—as our foreign aid program may have done at times—or they may serve as a stimulus to Indian policy, but they cannot serve as a substitute for Indian policy. The success of the Indian government in dealing with its own problems will determine its ability to survive as an effective working democracy—almost the only one in Asia today. The effect the United States has on this, in a positive sense, is inevitably peripheral, but it may be crucial if Indian democracy fails and a net is required.

I shall look at the problem of U.S. policy in a variety of fashions: first, in terms of India's relations with its neighbors, China and Pakistan especially; second, from the longer point of view of Indian economic development, with special attention to an aid program in terms of both the quantity of aid and the effective use of that aid to influence Indian policy; third, from the point of view of U.S. policy, especially on trade, both to assist economic growth directly and to influence Indian policy on economic development.

---

[5] See A. H. Cantril, *The Indian Perception of the Sino-Indian Border Clash,* Institute for International Social Research, Princeton: 1963.

INDIA'S RELATIONS WITH ITS NEIGHBORS

India's economic development and its political stability are threatened by its relations with its large neighbors. China offers a definite threat of invasion; with Pakistan there is the open wound of Kashmir, which has flared into open combat as well as increased internal tensions. Relations with both China and Pakistan threaten the existing secular democracy; at the same time the diversion of the large volume of resources for military purposes represents a potentially serious drain upon India's scarce economic resources. Unless India can be assured of a mostly peaceful environment and freedom from attack during this period of economic growth, it will have difficulty in achieving either the long-range results of economic growth or even any large-scale industrialization by the democratic political process.

The main issue dividing India and Pakistan is the Kashmir issue. A final peaceful solution will take time; and the effect of the deaths of Nehru and Shastri will be to lengthen that time. The United States can express the desirability of a solution and offer its good offices in arriving at one. I do not feel that a plebiscite is a means of arriving at such a solution at this stage. A plebiscite in Kashmir, which would be waged on the religious issue, can only worsen relations between Hindus and Moslems in both India and Pakistan, and it could result in the renewal of the rioting and blood baths of the partition period and in pushing India toward communalism. A plebiscite may be useful only to confirm a decision reached by other means. The United States should exert its influence against any military action to change the present situation. Meanwhile it should encourage a continual dialogue between the two countries and the conditions that permit this so that a peaceful solution may be worked out. But the Kashmir issue is a border problem that will exist for a longer time, and it obviously complicates U.S. relations with both countries.

Meanwhile, other problems exist in Indo-Pakistan relations, the solution of which with the experience of working together would create an improved atmosphere for resolving the Kashmir problem. An example of such a step is the long-negotiated agreement on the Indus-Sutlej waterworks, which the World Bank and the U.S. government backed, so that both countries can use the river waters jointly and equitably. There may be other possibilities of economic coöperation: for instance, steps to encourage coöperation in the production of the goods both India and Pakistan can produce economically, rather than uneconomic duplication of facilities in both countries, thereby taking advantage of economies from the larger market that would exist in both countries together. Both countries are major importers of machine products and capital goods or equipment for their development programs. In both countries internal markets for the output of such industries is small, and it is difficult to take advantage of economies of scale. To the extent that each country has excess capacity in the output of such a product or a comparative advantage in its output, it may be to its advantage to import it from its neighbor rather than getting it from the West as aid or otherwise. At the same time, this would permit the more effective use of the aid from third countries or even a reduction in the aid required. The United States might try to encourage greater trade between the two countries, in the hope that such trade would also contribute to greater political amity—at the same time recognizing that this need not necessarily follow—and permit more effective use of its own economic aid program to both countries.

On a technical level each country may be able to gain from the other. For example, many Indians have had experience in dealing with land and other problems in Pakistan or experience in India that would be useful in dealing with Pakistani problems. Similarly, Pakistani experience would be useful in India. Some technical exchange between the two countries would help in the economic development of both, and it would encourage further interchange on a technical level.

Granting the possibilities of these steps in a peaceful situation, the effects in the short run are likely to be minor, and would probably be unsuccessful in the immediate future because of the fighting over Kashmir in 1965. The United States will now have more difficulty trying to apportion its aid to both countries in a way that does not exacerbate their relations. This obviously raises questions with respect to continuing previous military aid programs to both countries, since this aid could further upset their relationship even though it might strengthen each country's ability to defend itself against a third power.

China is another potential threat to India. China is much stronger than Pakistan and is motivated by strong ideological and possibly nationalist grounds to try to increase its influence beyond its present borders. At the same time the threat of U.S. intervention to prevent Chinese expansion by open military attack is a major deterrent. The Soviet Union too has opportunity to exercise restraint upon China, although if there is an open split between them this is less likely. It also is a potential source of military assistance to India. In fact as relations between China and the Soviet Union get worse, thus reducing the Soviet potential for restraint on China, the potential for assistance to India increases. In case of deteriorating Sino-Soviet relations, the Soviet Union would be far more interested in maintaining an India free of Chinese control, and by its aid program to

India would antagonize China even more. The possibility of either Soviet restraining influence on China or, what appears to be more likely, increased Soviet economic and military assistance to India to contain China is a valid reason for India's policy of nonalignment between the United States and the Soviet Union. Such Soviet restraint on China or aid to India is not only valuable for India, but also for the United States since U.S. resources that might otherwise have to be available or committed for the defense or development of India can be used elsewhere. To the extent that the Soviet Union assists India in its economic or defense buildup, it may directly save U.S. economic and military aid expenditure; this is provided such Soviet assistance is in fact economic and does not at the same time increase the arms race between Pakistan and India. I think it can be assumed that the present Indian government does not wish to be dependent on Russian aid alone since such aid is so closely tied with shifting Russian relations with China and internal Communist interests. But neither does it wish to be tied only to U.S. aid. Indian acceptance of moderate Soviet aid should not be a major source of worry to the United States, nor should it serve as a major excuse for expanding U.S. aid to India to counter the effects of Russian aid.

To conserve India's scarce resources as much as possible, the Indian government must coördinate its economic development and its military program in a way that has not been forthcoming until now. Some competition is inevitable between defense and development, but the defense program should not be allowed to hinder an already lagging development effort. Finally, the expanded defense program should yield the maximum of external economies for development, both as a general stimulus and with application to specific projects.

In the area of defense itself, it is difficult for an outsider to say very much. It is clear that India in October, 1962, did not foresee the attack in the northeast frontier area as a Chinese effort to put pressure on India to gain its point in Ladakh. Indian defense planning in the future must try to evaluate defense in the light of various levels of potential foreign aggression ranging from internal subversion in the border states—Bhutan and Sikkim—to outright attacks of various strengths. It can then prepare alternative lines of defense against different foreign threats and the forces or policies to meet them. These alternatives should be tied in with India's economic resources and capabilities.

Although India's relations with its close neighbors other than Pakistan and China are not unfriendly, they could not be considered friendly prior to and during the 1962 crisis. This reflected an unwillingness by the smaller countries to antagonize China, but also there were genuine problems between a very large India and its smaller neighbors. There is little that the United States can do independently to assist in solutions of these problems, but it may be able to lend its good offices toward encouraging the negotiation of solutions if it is requested by both sides; and this would permit concentration of India's efforts upon the more threatened areas.

INDIAN ECONOMIC DEVELOPMENT: ECONOMIC POLICIES IN INDIA

I have already stressed both the importance of accelerating India's economic development for its continued political stability and the fact that politically some time still exists to achieve this acceleration. Until now U.S. policy with respect to Indian economic development has centered on the grant of aid—defined broadly to include gifts or loans—essentially to meet India's foreign exchange requirements for the Indian Plans. A large foreign exchange deficit is inevitable if India is to import the capital equipment for its new plants, and it will remain until the industrial plants become operative and India can expand its own exports. The deficit will probably increase still further in the Fourth Plan, and so India's requirements for economic assistance will increase rather than decrease. The United States has recognized this, and the aid it and the other countries have given has made the difference between the successes achieved by the Plans and complete failure. The United States has given the aid on the assumption that Indian economic planning and policies were of sufficiently high quality and were adequate for the requirements of development, and therefore American interest in the structure of the plans and in Indian policies for their implementation were relatively slight.

But the crisis of leadership that India has been passing through could not be foreseen, nor could the force of the political and ideological pressures upon the economic policy-making structure. These pressures have contributed to a set of policies favoring expenditure in the low-return sector of agriculture. Meanwhile, the political power to raise resources in the rural sector is very limited, so that there has been a flow of development resources from the nonagricultural to the agricultural sector. The development of capital-goods industries were favored at the expense of consumer goods. The development of import-saving industries were preferred to export industries or policies. Dispersed rather than concentrated investment was favored to satisfy the states. Reliance was placed on direct controls rather than on the use of the price system as an allocative mechanism. Each of these policies has its rationale, but carried to the level they have been they have contributed to misallocation of resources and the disappointing rate of growth of the Indian economy. It is here that the weight of American policies associated with the American aid program to India can make a difference in the short run. It can do this by seeking to redress the present bias in favor of the above set of policies and by encouraging a somewhat different set, which would again stimulate growth and induce structural change while allowing for the political framework within which the policies are formed. In the longer run the increase in the strength of industrial and urban political groups and the pressures and ideas generated within these groups would hopefully lead to a different political power structure and a set of policies that would continuously stimulate industrial growth. But within the next decade the political influence of the urban industrial groups is not likely to be much greater within the Congress than it is now, and it may be less.

In the field of Indian agriculture itself the United States cannot do very much directly by supplying a large volume of capital equipment. Indian agriculture does not need imported machinery. The United States in its own history has had a long experience with agricultural research in new inputs, new outputs, and new techniques. Although the products of that research may themselves not be directly usable in such a different environment, a technical assistance program to make the research technique known to Indian research institutions would be very desirable. This would aim at developing new inputs and outputs suitable to India's peculiar land and climatic conditions and encouraging their communication to the peasants. Where U.S. aid is associated with projects that create new agricultural inputs, such as fertilizer plants or new irrigation works, or in directly expanding final output of some farm production, the aid should be combined with efforts to influence policies. It should advocate a price policy that would encourage their use if they are inputs or expand their output if they are final products.

Apart from U.S. support in fields in which India is itself taking action, the United States may be able to influence the Indian government to take action in fields where its activity is otherwise unlikely. One of the areas where such influence might be used is the implementation and possible extension of land reforms to protect tenants in their rights by reducing the threat of eviction, and to encourage coöperative institutions to provide credit and other facilities to tenants. These would contribute to raising farm output and also have desirable social consequences. This is a politically sensitive area for U.S. policy to venture, but it is probably less sensitive than other types of land reform; and it also may be able to gain sufficient political strength to be achieved, since in many states it only means enforcing laws already in effect.

Another area in which U.S. influence might encourage action that would otherwise be unlikely is in the raising of direct taxes on the peasant in a progressive fashion. The political difficulties of both of these are great since land reform and agricultural taxation legislation are state prerogatives. It is therefore unlikely that any U.S. influence could be exerted directly: it would have to proceed slowly and by persuasion.

Perhaps the most important contribution the United States can make in the field of agriculture is to continue to supply agricultural surplus products, especially foodstuffs and cotton, as part of its aid program for an extended period of time. The supply of such surplus products makes possible, within limits, the continued growth of India's industrial labor force without being dependent upon greater agricultural output, since it provides food and clothing for this greater labor force at reasonably stable prices. Without this supply since 1961 food prices would have risen in the past few years even faster than they in fact did. Furthermore, such PL 480 aid can be used in conjunction with programs of employing surplus farm labor on labor-intensive construction programs for irrigation works, roads, and so on. Few if any such programs have been developed because of the administrative problems in operating them.

The supply of foodstuffs as aid has limits. First, on the U.S. side, the demands for these same surplus products may be rising from the commercial side, especially as the Communist countries are faced by their food supply crises and shortages. So there may be problems in maintaining the present levels of surplus food aid within the framework of present U.S. farm policies. Second, the supply of surplus food may create an unwillingness within India to take the perhaps politically unpalatable steps that are needed to increase agricultural output: to improve incentives, to stabilize prices with food stocks, to better allocate agricultural investment and expenditures for raising farm output directly; or to increase manufactured and other exports so that India might commercially import additional food and other farm products. The PL 480 program should be closely associated with India's own programs in both these areas. The surplus food program not only gives India food, it gives India time to develop its own programs and to increase its own efforts in the agricultural and export sectors. Both sectors require time for programs to be developed and implemented, in part because so little is known within Indian government circles about each. The agricultural sector is old with well-established techniques and institutions, but the export sector is relatively new, and it needs time to break into new markets and new areas. The supply of surplus foods under PL 480 is not a substitute for Indian programs; it is hopefully a condition for these, and American policy should aim at encouraging Indian programs.

In the nonagricultural sector, U.S. aid in the past has been largely concentrated in the economic overhead sectors, especially in transportation and power. Investment in these areas is important, since bottlenecks of both transport and power have had multiplying effects on the output of the entire industrial sector. But investment in these sectors is heavily capital intensive; there are some questions as to the economic returns on investment in either compared with alternatives. There are possibilities of using the rate structure for the outputs of both sectors more effectively than heretofore to allocate the demand for transport services and power. Nevertheless, continued expansion of these overhead sectors will be necessary. Aid that is given by the United States for both these sectors might be associated with suggestions to improve the rate structure in both.[6]

Relatively little U.S. aid has been given directly for industrial investments. Of what has been given the automobile industry has been a major recipient mainly for the production of trucks and jeeps. Much aid has been given to India in the form of nonproject assistance, and much of this has been for purchase of raw materials, spare parts, and so on; this has of course benefited the industrial sector. The U.S. Congress has been reluctant to vote assistance directly to government-owned plants, because of its ideological preference for private enterprise. I think this is a mistake in the Indian context, where the government may necessarily have to make such investments. I also think that aid to the automobile industry, possibly justifiable on defense grounds, has little other rationale. In India this industry has a small output scattered among several plants. Consequently, it is probably both a high cost and a low return industry with little export potential.

Aid should best be given to the new industries that do have a potential for export. India is developing its engineering and machinery industries; they are both import-saving and represent a potential area for expansion of exports. The United States, by providing modern equipment for such industries, would be contributing to reducing their costs, improving the quality of their output, and thus to expanding their export possibilities. It may be argued that greater Indian exports of such products would be competitive with U.S. exports. But at least for some time machinery exported by India is likely to be far simpler and less technical than that exported by the United States, so the competitive element would be very small. The contribution to India's foreign exchange earn-

[6] Recently both the World Bank and the United States Aid Mission, in conjunction with the Indian Government, have sponsored economic surveys of railroad transportation and electric power in India.

ings could be significant, thus reducing the total U.S. aid required for the economy. Aid to export industries would also provide an opportunity for American advice in developing policy and taking positive action to expand India's exports—by quality checks, price policy, servicing, and the like.

U.S. aid could have beneficial consequences in production of both consumer and intermediate goods for the rural population. Many of such goods are now being produced in limited quantities, such as fertilizer, small agricultural tools, and bicycles. The United States is a leader in the mass production and mass sale of such goods, and American techniques of production and marketing practices might be of value to Indian industrialists, who for a variety of reasons tend to be geared to the urban markets and a low volume, high unit profit policy.[7] The development of such industries would provide a basis for the creation of a much larger industrial labor force than at present and would encourage the flow of resources from agriculture to industry.

Aid might also be given to industrial research especially with respect to technology. Industrial research in India is a more or less neglected area. Indian producers borrow western technology because it comes with the machinery and because Indian technicians themselves are trained on such machinery. But there may be possibilities of developing less capital-intensive techniques for use in Indian factories, perhaps by modifications of existing western machines. The United States might sponsor research in such techniques and equipment either in American industrial laboratories for transmission to India or by setting up an industrial laboratory in India for such purposes. Such programs would directly stimulate the growth of the industrial sector at a more rapid rate and the structural change that has not yet occurred but that is necessary if Indian development is to be successful in the long run.

Indirectly, the United States can exert pressure to reduce the great administrative concentration of economic power as it is evidenced by the bewildering and delaying system of controls. The grounds for such a reduction are not ideological, they are now administrative and pragmatic. But the pressures—bureaucratic, ideologic, from private interests and the States—for the continuation of the present system are strong. Here again the United States, by associating its economic aid in general and for specific projects with conditions for the effective use of this aid, can exert its weight in favor of those groups within the government and the population who favor greater use of economic criteria for allocation pur-

[7] Walt W. Rostow has stressed the value of Sears-Roebuck techniques of selling in the underdeveloped countries, to begin to tap wider markets.

poses. The benefits, by releasing scarce administrative talent, are potentially great in stimulating the Indian decision-making mechanisms and so its economic growth.

Until now I have been discussing industries that produce civilian output. The requirements of defense call for expansion of India's defense production industries as well. Here the United States can help India build up its own defense production base by supplying technical expertise and equipment. The United States may also help to resolve India's dilemma over wanting to achieve self-sufficiency in defense production to maintain freedom in international policy and the pressure from its very scarce resources to develop only those defense industries that are economic—that is, producing goods at a cost not far from that of the cost of imports. These latter may be products that India would be able to export to other friendly countries so that expansion of this type of defense items at very high cost, compard ewith imports from other sources, tribute to India's economic development program. However, to produce defense items at very high cost, compared with imports from other sources, is wasteful of scarce resources and should, if possible, be avoided by India. Similarly, the production or even the importation of defense items that have a high prestige element but would be of negligible value in dealing with a real threat is undesirable economically. By encouraging defense planning that carefully evaluates potential threats, and by relating defense production planning to the economic resources and industrial capacities of the country, the United States can make a contribution in both respects. At the same time U.S. policy must realistically recognize that prestige and political elements are involved in defense.[8] It must weigh in its program of defense aid to India the conflicting claims of economics, defense, the influence of aid upon Pakistan and other countries in the area, and the political implications in India of the aid or its refusal.

A final technical area in which the United States can make a contribution is in encouraging research—both in India and in the United States—in the development of effective, low-cost techniques of birth control that are usable within the present Indian environment. It should also encourage the Indian government to place greater priority in its plans for communicating the results of research and for developing and testing operational incentives that will encourage the rural population to limit family size. This latter part of any program obviously can be handled only within India, since it calls for full awareness of the economic, religious, and cultural aspects of population control. It is possible

[8] This is of course true in India as well as the United States.

of course, that American communication techniques and tools would be useful in this area, but Indian experts are the best judge of this.

In summary, these suggestions with respect to the U.S. aid program for India add up to a general policy somewhat different from that of the past. They require in effect that the United States become far more deeply involved in Indian planning than it has heretofore. The United States must have its own conception of what are desirable goals for the Indian economy and know the desirable methods of achieving those goals. With those goals it can participate as actively as possible in assisting the Indians in preparing its plans.

It is most important that U.S. goals in India not be framed in a narrowly ideological fashion, nor that any advice to India be presented on such grounds. In my opinion the ideological interest of the United States is well satisfied by the fact that India is a reasonably effective working democracy characterized by free elections and an independent court system, and that it has a high degree of free speech, freedom of the press, and freedom of assembly. Whether within that framework India wishes to pursue its policy of economic development with a greater or lesser amount of state action or private enterprise is a decision for India to make; also, it is an area in which India—like Japan in a comparable period—may need to make greater use of government operations than the United States required in a similar period or feels is desirable now.

If we accept both the desirability of increased U.S. involvement in Indian planning and the importance of the nonideological element in that advice, the United States must have and use the technical expertise on Indian economic affairs that can make and support suggestions on plans and specific policies. Without such detailed support on specific policies, the suggestions will not be received, or can be dismissed, and rightfully so. With support, the suggestions can become policy issues within the Indian government, and they will then be discussed on their merits, as both political and technical matters. At the same time the reduction of the ideological element in India's policy making would improve the possibility of successful discussion on these issues.

To develop such technical expertise on the American side requires that the appropriate American agencies within India be staffed by a reasonably large number of qualified personnel who are capable of understanding the functioning of the Indian economy and are given the time to devote themselves to analysis of problems and suggestions for policies to deal with them. This implies that these American personnel will have the opportunity and will themselves be able to develop a relationship of mutual trust and confidence with the Indian administrators and economists, and if need be, Indian political leaders. Suggestions emanating from American sources will then be both offered and received in a spirit of mutual respect and faith in good intentions as part of a discussion on matters of great mutual interest.[9] The Indian officials can use the support from American aid programs to support their own policies of a similar viewpoint, and they will be able to do so if this support is offered as an exchange between equals and not as an American policy based only on ideological grounds or in exchange for aid, and unsupported by evidence.[10]

Finally, the United States should be prepared to use its aid programs in India to give aid in support of one set of projects and to withhold or reduce aid to oppose another set when it feels strongly about the choice of the projects and the policies associated with them. The United States should have the staff on hand to support its choices among alternatives on economic and not ideological grounds. Until now it has withdrawn from this exercise of influence. It must of course exercise this power carefully, perhaps as a last alternative, and with respect for Indian susceptibilities. But failure to exercise this power with flexibility and with courtesy can lead to a neglect of both the Indian and the American interest in the Indian aid program and Indian development.

It may be urged that, if the United States becomes more deeply involved in Indian planning, it in effect becomes committed to support those plans to the full extent necessary if programs are agreed upon. Given the requirement of annual congressional aid appropriations, this is not a commitment that could be made. The point is a valid one. The argument against it is that the United States, the other western countries, and the International Bank are committed to the success of Indian development, both because of the heavy past investment in it and because the consequences of failure could be politically disastrous. By failing to become more deeply involved in the process of planning that

---

[9] This would be a relationship closer to that developed by private personnel of the Ford Foundation. Although an employee of the United States has certain disadvantages from his official relationship, it should be possible to develop mutual interest and confidence if a sufficient number of able persons are sent to India and if they are given both the time and the resposibility to develop such a relationship by involving themselves deeply in Indian planning.

[10] Although this policy of greater interest in India's plans and implementation policies is presented in terms of the United States, there is no reason why this should be so limited. The World Bank prepares general reports on problems of the Indian economy for the information of the government of India and hopefully as a guide to policies for it; and the consortium of aid-giving countries provides an instrument for bringing together other countries—excluding Communist countries—with an interest in India's growth. Such an international source of policy influence might well prove more effective than if advice came only from the United States. Communist advice might also prove useful.

development, the assisting countries are increasing the chances of the very failure of the development that they are aiding and that they wish to succeed. The implicit commitment has been largely made. This is in my judgment a valid reason for becoming more deeply involved in the planning process and for supplying aid in appropriate amounts so long as the prospect of success seems reasonable and the political environment is considered favorable. This answer of course does not deal with the very hard question of how the American Congress can be persuaded to support such a commitment every year. But it requires that Congress and the public be better persuaded of American interests and goals in India and of the steps required to achieve those goals.

INDIAN ECONOMIC DEVELOPMENT: POLICIES IN THE UNITED STATES OR IN THIRD COUNTRIES

Apart from the aid program taking place in India, the United States has other policy instruments, external to India itself, that can influence Indian development. The most important of these is trade policy.[11] India does need to increase its exports, especially of the new products coming from its factories, and the possibility of increasing the traditional exports is quite narrow though it has not been exhausted. The developed countries are potential markets for some of these newer products and so are other countries now in the process of development. In the United States and the other developed countries, the political pressures against reducing trade barriers to Indian goods are high; but given the quality of Indian goods and the prices, the problem has not as yet become serious. At the same time, if India is ever to repay its past loans and not require further aid, those barriers will have to be reduced. To the extent that tariffs exist on traditional raw material exports, such as jute and tea, they might well be immediately reduced.[12]

At present if the United States reduced tariffs generally on the new manufactured goods produced by India, India would have little benefit. Those goods are in many cases simply not competitive in the United States because of their high price and low quality, and the benefit would go to developed countries producing similar goods. India and the under-

---

[11] On the use of trade as a policy tool see J. R. Schlesinger, "Strategic Leverage from Aid and Trade," *National Security: Political, Military, and Economic Strategies in the Decade Ahead,* eds. D. M. Abshire and R. V. Allen (New York: Frederick A. Praeger, 1963).

[12] To the extent that quotas could be reduced on exports of cotton textiles to the United States and other developed countries this would be desirable, but there are great political difficulties.

developed countries have asked for special preferences for the exports of their manufactured goods. Although the theoretical arguments against such special preferences may be valid, their effects in the real world could be beneficial and the subject should be treated pragmatically. Consideration might be given to a system of preferences on a sliding scale basis, so that over a period of, let us say, five years, they would be eliminated. At the same time, where such preferences are given they might be associated with a *quid pro quo* that the underdeveloped countries develop and begin implementing programs to increase their exports by reducing costs, improving quality, providing service facilities, and so on. This would make it possible for the Indians to compete more effectively than at present in the underdeveloped countries, where the barriers are more of India's own making, since those countries often do not have existing or growing industries that they are protecting.

At this stage the United States should begin preparing its long-range trade policies toward India and other underdeveloped countries. It would gain economically if it were eventually able to purchase low-cost Indian manufactured imports in exchange for surplus American grains and relatively advanced American machinery and services of various types. The effects on specific U.S. industries could be minimized by policies to accelerate the rate of American economic growth to approach closer to full employment and to develop special programs specifically for affected industries and individuals. If this general approach is acceptable, it means that the United States must begin thinking in some detail of its import policies on specific Indian products, as well as products from other developing countries. This would permit the United States during this decade to carry on the necessary negotiations to permit greater imports from the developing countries, including India, into the markets of the developed countries, including the United States. It will also make it possible to devise phased programs that minimize specific industrial and regional costs of such imports so far as industries and areas of the United States are concerned. But because these changed import policies will involve costs to groups in the United States, they should be associated with bargaining to change India's trade policies. Those that are inimical to India's own development and thereby involve aid-giving countries in higher aid expenditures, or those that may be prejudicial to legitimate American interests, India might be encouraged by this means to give up. Any such policies and programs are obviously deeply involved in U.S. politics. This may create short-term difficulties or problems on specific products, but the whole trend of U.S. policy of protection since 1934, as well as the experience with Japan—especially after the war—indicate both that U.S. protective policies are amenable to change in the relatively short run and that new or recovering producing countries can find markets in both developed and underdeveloped countries.

CONCLUSION

In this chapter I have not attempted a detailed analysis of U.S. interest and policies in India. Such an analysis would require a separate study at least as long as this one. I have drawn out some implications for U.S. policy of the earlier analysis of the trends in the Indian political system, in policies on development, and in the Indian development. Accepting the U.S. interest in a stable democratic political system in India, which in turn implies an interest in a successful achievement of economic growth, implies certain things for our own policies. First, the United States can assist India to buy the time necessary for its economic plans to come to fruition. By supplying foreign exchange to India in conjunction with other countries, and by supplying surplus agricultural products—largely food grains—economic aid programs have made it possible for India to achieve some growth without undue pressure on Indian consumption and with rather slowly rising prices. As yet the United States has no trade policy with respect to new Indian exports; but this has not been important since the exports have not been coming, although the demands for relaxation of barriers against them have been.

I think the United States is fortunate that India has the time to develop adequate policies geared toward continued economic growth. The pressure of internal politics reflecting conflicts among important political groups in the Congress party, economic and bureaucratic vested interests, the state governments and various ideologies combined with the weakening of India's heretofore strong political leadership make it difficult to develop those policies and mechanisms that lead to continued development and the changes in the structure of the economy that must go with development. This has contributed to the present difficulties in the economy. Therefore, in spite of the high quality of India's economists and officials, the United States must play a more active role than heretofore in influencing Indian plans and implementation policies on development. It must try to use its instruments of aid and trade to stimulate those policies it thinks desirable. Within the Indian administrative structure there are differences on specific policies, and American support for one set as opposed to another can make a difference. It is most important that the reasons for our advocacy be not ideological nor presented on ideological grounds. This means that the United States must have in New Delhi personnel on both the economic and the political levels capable of analyzing Indian plans and policies and suggesting specific changes.

They should be ready either to provide technical support for these changes or to suggest research. And they should have the opportunity of advocating these changes with their supporting arguments to Indian economists and political leaders. Finally, there needs to be a willingness on the American part to use aid and trade programs—both by changes in total aid allocated and by distribution of aid among specific sectors or projects—to support the recommendations made by its qualified representatives. This does mean a deeper involvement in Indian policy formation than heretofore. It also requires the United States to recognize that, even with this involvement, American influence in Indian policy-making will be necessarily limited and should therefore be used both modestly and carefully if it is not to be wasted and result in antagonism rather than mutual benefit. Failure to use that influence will not merely lead to waste of our contribution; far more important, it will jeopardize the success of India's entire development program, and thus contribute to undermining the common interests that both the present Indian government and the government of the United States share.

PART VII

Epilogue

"Developing a country under democracy is like trying to play poker without bluffing."

B. Taper, "Letter from Caracas," *The New Yorker,* March 6, 1965.

In an underdeveloped country, experience proves that the important thing is not that three hundred people form a plan and decide upon carrying out, but that the whole people plan and decide even if it takes them twice or three times as long. . . . People must know where they are going, and why.

F. Fanon, *The Wretched of the Earth*

[It] is to be recognized that in Soviet Russia economic decisions most frequently are a function of political necessities or desirabilities of the Soviet government. The mechanics of the exercise of dictatorial power make for the primacy of politics over economics, the tenets of the materialist conception of history to the contrary notwithstanding.

A. Gerschenkron, "The Study of The Soviet Economy in the USA," *Survey*

14

Some General Applications

In all underdeveloped countries there is an interest in economic development, and the governments are deeply involved in planning for it or implementing plans. The goals and processes of economic development are closely involved in the politics of the country, and they raise issues that affect the varying interests of different social groups and conflicting ideologies. In India we have looked at the rural caste and urban class structures as social systems within which power and influence are exerted. In the African countries south of the Sahara the caste system does not function, but perhaps an analysis of tribal structure and the changing relationships between tribal structure and economic and political power in rural areas would provide a method of analyzing political change and economic policy formation in many of the new lands of that continent. In urban African areas, class factors may interrelate with tribal factors to affect power and ideology, and in both rural and urban areas the racial problems of white and black would introduce further complexities.

In some of the Latin American countries a fruitful framework for policy analysis might be provided by the social distinctions between groups with Indian background and those with Spanish background. In turn, this might, depending upon the country, influence or determine the structure of landownership and of political power. These differences might also serve as a basis of power in the cities, or they might be overshadowed by straight economic conflicts. It is of interest that several economists have suggested looking at inflation in some of these countries, such as Chile, as a political phenomenon—as a method of satisfying powerful class and economic groups in the country even if it harms the less important groups. If this is correct, analysis of some of these other countries, in terms of the framework outlined in this book, would be useful.

The problems of growth in India are not peculiar to that country. Historically some or all the problems associated with increasing agricultural output, shifting resources from the rural sector to the industrial sector, improving incentives for agricultural and industrial output both, reconciling demands for greater personal and regional equality, taking steps to accumulate capital and allocate investments among different groups of industries and areas—have been major economic and political issues in such diverse countries as England, Japan, and the U.S.S.R. in their development processes. These same problems exist to some extent in all the underdeveloped countries. Their solutions are influenced by the ideologies present in those countries and by the strength of various political groups, as much as by the economists' advice. One of the factors that the economist, whether native to a country or foreign, must consider when he advocates policies is the effects of those policies on the various interested groups in the society and their potential gains or losses. The foreign adviser can do this but rarely for obvious reasons, but he should know his advice will be considered by his audience within a political context; he should be aware of this fact in framing his advice.

This study raises another series of questions relevant to many underdeveloped countries—the relationship between democracy and economic development. The case of India is directly relevant to this question. First, from the Indian case it is clear that a policy of economic development planning and its implementation is not incompatible with democracy. Second, democracy gives political weight to numbers, and in an underdeveloped country almost by definition the largest numbers—and perhaps the major political power—are in the rural sector of society. The political influence of the rural sector of society makes it politically difficult either to carry out a massive program of industrial development or to divert resources from agriculture to industry. However, policies to stimulate investment and growth in both sectors are quite possible.

The Indian experience, at least for the decade of the 1950's, shows what a democratic government with a program and effective leadership, of the type supplied by a unified Congress party under Prime Minister Nehru, can do: with massive economic aid, they achieved a politically satisfactory rate of economic growth. Nehru was capable of holding together a coalition that was able to govern effectively and carry through policies desirable for growth in a wide variety of fields. He was, during that period, an effective "reform-monger." The difficulties that India faced in development after 1961 arose partly from natural causes, such as the poor monsoons. Equally, they arose from the crisis within the Congress party, a crisis of ideas and leadership. The ideas that served successfully in the 1950's, which largely derived from the 1930's, were no longer satisfactory as a basis of policy; but new ideas had not been developed. The leadership, for reasons of age and health, no longer proved capable of generating new ideas to serve as a basis for policy. The Prime Minister no longer led the party, he held it together. At the same time a conflict arose between the old ideas and the interests of powerful new groups within the party. All of these, combined with several poor harvests, have contributed to a period of policy cross purposes, political stagnation and economic stagnation, since government policy is such a crucial element in Indian economic decision-making. What is required now is another set of ideas and a new willingness to experiment: India needs to develop new economic policies to suit the new circumstances, she needs to achieve a political coalition to support those policies and begin to implement them, and she needs to press forward with those policies that appear successful and discard those that fail. The questions in India are whether these new ideas will arise, and if so whether the political constellation will permit them to be translated into policies.

Does a dictatorship necessarily have advantages over a democracy in carrying out development? A dictatorship per se does not. There is no reason to believe that a dictatorship need necessarily represent groups in favor of economic development; the reverse may just as easily be true. The assumption that a dictatorship is a single will is also false; it too, like a democratic political party, represents a mixture of forces advocating conflicting policies that emerge compromised, but conflict exists behind the scenes. A dictatorship that represents a group committed to economic development and has the power to carry out its policies and the popular enthusiasm for support of those policies, even if onerous, may be able to experiment with a variety of policies without fear of overthrow and may achieve spectacular gains—at least in the short run. But such a dictatorship can also achieve spectacular failures. The evidence of declining economic well-being in Indonesia testifies to the failure of a dictatorship that has not been interested or able to carry out reasonably rational economic policies. Similarly, even within the Communist bloc, Yugoslavia, Hungary, and Poland adopted economic policies that were unsuited to the resources of the country, and they were changed only after drastic political upheavals and policy reversals. Even in the case of spectacular gains associated in the short run with a dictatorship, a question arises as to the long run. This is true especially if those gains are the result of imposed actions or actions that offend many. Turkey's problems since Ataturk may reflect this conflict, as might those of China and Russia in the area of increasing agricultural output. In the longer run the policies adopted by democratic action may be accepted, whereas those imposed from above may be abandoned. The difficulties and crises of development are more obvious in a democracy, of course, since they are discussed freely. In a dictatorship this is absent, so it can create the illusion of "no problems" until a day of reckoning arrives.

The question is not the form of government, but the quality of its leadership, its ideas and policies, and its success in creating political support for those ideas and implementing them by policy. For a country with no background for democracy, with no experience in creating a politically effective government, and with an ineffective bureaucracy, democratic methods may not be suitable for economic growth. For a country with some of that background and experience democratic methods may be possible, and India has proved it so even with its great and ever-present problems.

The United States may find it necessary or desirable to support a country governed either as a democracy or a dictatorship, given its aims and those of the country and its understanding of the political and economic forces and pressures within that country. It would be ideologically preferable to support a democratic form of government, but such choices are rarely possible. In the case of India the United States has the relatively rare situation of being able to exercise this preference for democracy in a country where economic development is associated with democracy, and it has the choice of being able to exercise that support largely by economic assistance and policies. But in the case of India, and in other cases of countries deeply committed to economic development as a goal—although one of the major contributions it can make is direct economic assistance, both by providing the necessary imports and by giving the aided country time—the support the United States can give by assisting the leadership of the country in developing new ideas and policies, in looking at their problems in an operationally significant fashion, and in developing effective reform-mongering coalitions to carry out these policies may be even more significant than the aid itself.

POSTSCRIPT
SOME AFTERTHOUGHTS ON DEVELOPMENTS IN INDIA SINCE 1965

"If the Congress . . . did not retain the requisite purity and strength [it] would find all power slipping from [its] hands. [It] could not hope to maintain it with the help of the bayonet like the British. All [its] power came from the people, who were the real masters, though [the Congress] might not realize it at the moment." —M. Gandhi

The original edition of this book was largely based on Indian developments through 1964. Since then India has passed through two years that have been described as the worst years since Independence, and has had a fourth national election. The purpose of these notes is not to bring the original edition of the book up to date, but to explore whether the ideas presented there have or have not held up, and to indicate some of the implications of the events since 1965 for the future.[1] These developments also raise new issues for U.S. policies toward India, and it will be of interest to explore the validity of the comments on those policies presented in the earlier edition.

The afterthoughts are not based on extended research into the recent problems of the Indian economy. Since 1964 I have neither carried on research into those problems, nor have I had an opportunity to be in India for more than several short visits. However, I have followed the events in India from a distance, and have often discussed those developments with both Indian and American friends. Therefore, these comments are presented as my opinions, with only a minimum number of supporting references or statistics. The topics are presented in the order in which they are treated in the text—political developments, economic trends, the interrelation of these with the Indian government's policies on economic issues, and, finally, the implications for U.S. Government policies.

The most important political event in the last three years is the 1967 election. Although it is far too early to make a proper assessment of the results and evaluate their consequences, it is clear that this election is a watershed in Indian politics since 1947. In the three elections from 1952-1962, the Congress Party won only about 45 per cent of the popular vote, although it at all times held over 350—or about 75 per cent—of the approximately 500 seats in the Lok Sabha, and with minor exceptions kept control

[1] These thoughts have benefited greatly from information supplied and comments by RAND colleagues Alan Carlin and William A. Johnson and by Professor M. G. Mueller of the University of California, Los Angeles, Leo Rose of the University of California, Berkeley, and Wayne Wilcox of Columbia University. None of these, however, are responsible for any opinions in this chapter. This postscript is based on information available March 1, 1967 before all the 1967 election returns were known or new governments chosen at the Center and in many states.

of the sixteen state governments. After the election of 1967, in which the Congress won 39 per cent of the vote, of 521 seats contested in the Lok Sabha, the Congress Party majority will be reduced to less than 50, although this may be increased somewhat if independent nominees, who are sympathetic to the Congress Party, decide to support it. Furthermore, and perhaps more important, the Congress Party failed to win an absolute majority in the legislatures of eight states—the entire northern tier of states, excluding only Gujerat and the new state of Haryana, but including Orissa, and the two states of the far south, Madras and Kerala. In three of these states the opposition parties have apparently won clear majorities and will form governments. In the five others, if the Congress proves able to form governments, it will be either because dissident Congressmen return to the party or a coalition is set up in which the Congress Party is the stronger element, but subject to overthrow.

The victories of the opposition parties do not so much reflect either agreement with their policies or the creation of a national opposition at this time. In the South the gainers have been the communal anti-Hindi DMK party in Madras and the Communist Party in Kerala; in the North the gainers have been the Swatantra Party, which ideologically is a democratic party that favors a free economy and the Jan Sangh, the Hindi communal party. These parties only agree ideologically in their opposition to the Congress, but in many states they succeeded in forming an electoral coalition that cost the Congress its previous majority. The losses of the Congress are probably due to two major factors, as well as numerous local issues. The Congress, after twenty years of power, had become "fat" and subject to criticism of widespread corruption and "bossism;" the economic disasters of 1965 and 1966 were blamed upon the Congress government and party, in the same way that the party had reaped the credit for the economic growth that had occurred until 1962. The failure to fulfill the expectations built up by the first three Five Year Plans and the past successes redounded unfavorably upon the Congress in 1967. And these factors combined to defeat the Congress in such states as Madras, where its record had been good, and in Orissa, where charges of corruption were rife.

The results of these elections mean the end of an era of complete Congress dominance, during which the political decisions were made within the Congress as a result of the interplay of personalities and factions, on both the national and state levels. Hereafter, the Congress Party will be the leading party, still with a national majority in the Lok Sabha, but one that will be far more influenced than previously by the actions of the opposition parties, and by minority factions within its own ranks. If the Congress Party suffers defections in the Lok Sabha, its apparent majority is small enough so that it can be overthrown on a national level; and on the state level it will have to deal with opposition governments in several states, or with coalitions in others. As a result, the process of integrating policies on the national and state levels will be harder than in the past. This latter problem will be even more difficult because there is no single major opposition party united on policies; rather the opposition parties disagree more sharply among themselves than they do with the Congress. For example, the disagreements between the Communist Party, likely to govern Kerala, and the Swatantra, probable governing party in Orissa, far exceed the differences of either with the Congress on economic and political policies (although all three are secular parties); the differences between the DMK in Madras and the Jan Sangh in the North are similarly far stronger on issues of language and communal relations than the differences of either with the Congress.

These differences and the greater importance of minority elements in the Congress will probably strengthen the power of state and regional leaders at the expense of the central government. The state opposition leaders will be attempting to show their independence of the Congress, and the national Congress will attempt to bolster its defeated state parties. Such a devolution of power from the center to the states was already occurring prior to the election within the Congress Party. The state and regional leaders within the Congress, organized informally in the so-called "Syndicate," simply ratified the choice of Shastri as Nehru's successor. However, when Shastri died unexpectedly they chose Indira Gandhi as his successor. They showed their political strength prior to the election in the criticism by the leader of the Congress Party of the government's economic policies; by the veto various state leaders exerted over Mrs. Gandhi's efforts to reshuffle her cabinet in late 1966; and by their veto over the choices of Congress Party candidates for the Lok Sabha elections for 1967.

This new relationship between a Prime Minister and the Congress Party leaders was described as one of consensus, and one commentator wrote:

> Unless [Mrs. Gandhi] attunes her manner of functioning to the realities of power in the Congress she cannot expect to remain in office long; no Congress Prime Minister today can. Effective power at the centre in the Congress is, and will remain, in the hands of men who control State Congress organizations. . . . The Prime Minister . . . will have to have the support of the group or at least the more powerful members of it. In other words, he will have to rule by consensus, by compromise, by bargaining. Of course, as happened in the Shastri regime, consensus will slow down the pace of working of the government, give the impression of indecision and vacillation. . . . These consequences may not be to the liking of the Prime Minister and some of them may not be to the best interests of the country either but all the same they are unavoidable if the Prime Minister is to carry the party with him as he must.[2]

[2] "A Lesson in Congress Politics," *Economic and Political Weekly,* I, No. 13 (November 12, 1966), pp. 520-521.

This rising power of the regional leadership contributed to the decision to divide the State of Punjab into two states along communal lines, and to the position taken by the Congress Party in Madras in favor of abolishing the land tax, which would benefit peasant landholders against other groups in society.

With the greater strength and victories of communal parties in the north and south, it is likely that such communal issues as language, cow slaughter, and others will become more rather than less important and the difficulty of compromising these issues will become greater, since those parties in favor of extreme solutions in various sections of the country will be stronger. It is possible also that the Congress Party, in an effort to weaken the communal parties, will itself try to take a less secular and more communal stand on some of these issues. If this does occur the political difficulties of governing India will increase as a result of the elections.

However there are several major qualifications to this somewhat depressing picture. The Congress Party still has a significant legislative majority and is able to govern the country and most of the states. Also the movement from single party government to a broader basis was inevitable in a democracy, and is desirable. The surprise was that it occurred in this 1967 election rather than at a later time as the first edition anticipated. This surprise was further heightened by the massive Congress defeat in Madras. The defeats in Kerala and Orissa were neither surprising nor totally unwelcome. In other states Congress losses do not appear to be very large at this time, and in several states the party did better than expected. Thus the transition from single party to multi-party government is occurring under rather favorable conditions in which the Congress still provides a framework of political unity throughout the country, but it has received a strong warning to improve its performance if it is to continue in power.

The greatly reduced majority of the Congress Party was accompanied by the election defeats of major regional leaders who were members of the "Syndicate." As a result, their prestige and position in the Congress Party, as well as their power, may decline. This would strengthen the position of the Prime Minister at the expense of the regional leaders within the Congress Party. To the extent also that the losses of the Congress reflect voter impatience with Congress sloth and presumed corruption, they may also contribute to the streamlining of the party, to a greater receptiveness to new and younger party leaders and their ideas, and to closer contact with the general public. If an effecive Prime Minister is chosen who is able to take the lead in reviving the Congress Party, he or she might be in a better position than in the recent past to resist the more extreme regional pressures. The fact that the Congress Party still has a small majority in the Lok Sabha provides some time for such a display of leadership in the reactivation of the party.

Another effect that can prove beneficial is that state power may transform some of the regional opposition parties into a national opposition party with some hope of actually forming an alternative government, as well as an electoral opposition, to the Congress. The Communist parties are not interested in the maintenance of democracy; but the Swatantra Party working together with the DMK and the Jan Sangh could lead to a new right wing party with a much broader base than any one of the three, and one in which the secularism of the Swantantra would diminish the communalism of the other two. The strong anti-Hindi victory in Madras could also moderate the pro-Hindi agitation.

But at this time these favorable possibilities are to some extent hopes revolving around the reversal of past trends; if recent trends continue, the results are likely to be a period of difficulty for the national government. In turn this could lead to a situation of political instability that would seriously threaten the future of democracy in India if the Congress does not succeed in improving its functioning after its losses and if a responsible opposition does not develop in the states.[3] The next few years will be a crucial, and possibly shaky, period of political transition in Indian political history since Independence—as the country moves from single party to multi-party democracy.

The importance of economic factors in contributing to this change is clear. The weaknesses of, and dissatisfaction with, the Congress became strikingly apparent in the face of the extremely poor crop of 1965 and the disappointing crop of 1966; the greatly increased demands of defense upon the economy, caused by the fighting with China and Pakistan; the cut-off in much of India's foreign aid from October 1965 to early 1966 following the fighting with Pakistan; the difficulties in expanding exports; and the reduced rate of industrial growth. During the Third Five-Year Plan period, unemployment is estimated to have risen by two or three million people; in the three years from 1963 to 1966 the general wholesale price index rose by 36 per cent, and food prices by 48 per cent (a greater rate of increase than in the previous ten years).[4]

The pressure of these events and requirements has led to some changes in Indian government policies since 1964. In agriculture, the government has

[3] A serious and astute reporter of Indian politics, Mr. Neville Maxwell of the *London Times* has flatly predicted that the 1967 election will be the last free election of India because of the devolution of power to regional leaders within the Congress. (See *The Times,* January 26 and 27, 1967, p. 11 in each.) At this stage I think he is too pessimistic. The 1967 election has shown India's attachment to democracy. There are favorable factors at work and India has survived other problems in the past. But I think the probability of the threat has risen over the past two years.

[4] Government of India Planning Commission, *Fourth Five Year Plan: A Draft Outline* (New Delhi, 1966), pp. 4-5; 106.

taken steps to attract private investment into the fertilizer output, to accelerate development and application of improved seeds and new crop varieties, and to concentrate rural development efforts in the most promising regions. In the industrial field, some of the controls upon entry of new firms into industry have been removed or simplified, some price controls have been removed, and the rupee was devalued. In the more general area of the Fourth Plan itself, although lip service has been given to a large Plan in money terms for political reasons, in fact the Plan in real terms will show only a modest increase. This is the result of the devaluation and the decision that in the first year the Plan will emphasize maximizing output from existing capacity, rather than building new capacity. It appears, too, that administrative talent has been shifted to deal with the problem of population growth, and that a major effort is being made to popularize the use of the intra-uterine device.

Although the Congress government was able to follow up its decisions effectively in some of these areas, in other areas it proved unable to do so because of its weaknesses. Devaluation aroused such a storm of criticism, both on ideological and nationalist grounds and within and without the Congress, that the government could not effectively carry forward the necessary follow-up steps to reduce deficit financing by the central and state governments and to tighten credit to private industry to prevent general price rises, apart from those arising from the devaluation itself. Prices continued to rise rapidly, in part because of the poor 1966 crop in many areas, and this was blamed on the devaluation.[5] Similarly, the decisions to make private foreign investment in the fertilizer industry more attractive by reducing controls on distribution and prices of fertilizer were heavily criticized on ideological and nationalistic grounds, and their implementation proved difficult. The beneficial economic effects of less politically delicate policies, such as an improved seed program, or an expanded birth control program, will not be felt rapidly. They will yield little, if any, political gains in the short run of a year or two, although in the longer run of five years or more, they may, as many observers believe, yield great economic gains, and perhaps political gains as well.

The events of the past two years have, if anything, made more obvious the difficulties of carrying out economic policies in the face of the growing powers of the states, and the ideological conflicts within the Congress, and throughout the country. Some of these were clear before the election. In the more general area of economic development, the growing power of the states made the reconciliation of the pressures from the states a key problem of planning. Each state wishes to increase the number of projects in that state while minimizing the taxes it will levy within the state by raising the share of taxes it gets from the central revenues. The central government tries to avoid an uneconomic dispersal of projects among the states. Inevitably the result in

[5] "Has Devaluation Failed?" *Capital,* November 24, 1966, pp. 929-931.

the Plan is a political compromise, but with the shift of power to the states it will be more difficult for the Center to resist the regional pressures brought upon it.

In the field of agricultural policy specifically, an example of the growing power of the states has been the continued narrowing of the free market area for the distribution of domestically produced food grains. India has for many years had a zonal distribution system for wheat and rice, which meant in effect that marketing of food crops was more or less free only within "a zone." Originally a zone included several nearby states, some exporting grain and others importing it, in order to achieve a regional balance. Within the past few years, however, the zone has been narrowed to the individual state, and marketing outside of that state must be done through the state government.

This narrowing of the market reflects an effort to achieve state self-sufficiency in grains by retaining as much as possible of the surplus output above subsistence within the state. Although this builds grain stocks in the rural areas and keeps prices low in the urban areas of a single surplus state, it aggravates the national food problem. The demand for state self-sufficiency leads in turn to a demand that every state, regardless of its natural resources and ability to expand its food output, receive equal treatment in allocation of investment resources for agriculture, even though the states differ widely in their capacity to raise agricultural output. It also results in a delayed movement of grain from surplus to shortage areas. Thus, there is a greater demand than would otherwise be required by the central government for food imports and aid, since foreign food imports are controlled by the central government and can be sent directly from the ports to the shortage areas. The effect of the limitation of the market to a single state is thus both to reduce the Indian output of food grains by poor allocation of resources and to increase the demands for food imports and aid beyond what they would otherwise be.[6]

The results of the elections are likely to make it even more difficult than in the past to moderate the demands of the states upon the Center. State governments run by parties in opposition to the Congress will be even less willing than previous Congress governments to seem to yield control over state policies to the Center. In fact they are likely to demand even more

[6] This is not to say that in the face of the great shortages of food grains in India in the past few years that distribution of food should be free. Under the conditions of these last two years a completely free market in food grains would probably have led to rapid bidding up of food prices and hoarding of grain by the wealthy. The introduction of food rationing in the larger cities has been effective in preventing starvation in those cities. But the argument above is that the restriction of the market to a small geographic area acts to reduce the supplies of foodstuffs below what they would otherwise be in deficit areas. If and when supplies of foodstuffs on a national basis increase sufficiently to meet India's normal requirements, it would be desirable to move to a free market in the sale of foodstuffs.

from the Center. The opposition governments and the state Congress parties will put pressure on the Center to adapt allocations and national policies to state demands. Until now the national government has succeeded in avoiding locating a steel plant in the south, but after the riots in Andhra for such a plant, and the defeat in Madras, this may no longer be possible.[7] (A contrary influence may operate, however, if the Center feels that it does not have to pay as much attention to opposition parties, or to weaker state Congress parties, and can therefore allocate resources on a national basis, or put greater pressure upon the states to accede to national policies than previously.)

The role of economic ideology in influencing national policy prior to the election was clear in the debate over devaluation. This debate clarified the relationship that had existed among three elements of government policy—an overvalued rupee, the system of exchange control, and the public sector investment program. Since this relationship was omitted in the first edition it will be worth spelling out in some detail.

The foreign exchange available to India is the sum of the amounts acquired through commercial exports and invisible earnings such as tourist receipts, foreign aid both of a non-project and of a specifically project type, and private investment. The allocation of all foreign exchange is done by the government. Prior to devaluation, in the allocation of the foreign exchange proceeds from aid and commerce, government projects received first priority, in part because control was exercised by the government, and in part because the public-sector projects were believed to be of first importance; so-called "maintenance imports" consisting of raw materials, spares, and so on, for existing plants received second priority; and additional private investment in new capacity received third priority. In fact, foreign exchange for private investment had to be self-generated, in the sense that such private investment would be permitted only if the foreign exchange were supplied by private collaborators.

The government, like other claimants for foreign exchange, had to raise an equivalent rupee sum. The artificially low price of foreign exchange made the rupee cost of new government investment lower than it would have been with a more realistic rate. Given the fact of a limited budget for development and the difficulty of raising additional tax resources, government investment projects were able to command a larger proportion of the available foreign exchange than would otherwise be available. In effect, exchange control with an artificially overvalued rupee operated as a tax on the economy that was politically more acceptable as a method of raising resources for government investment than was a general increase in taxes to finance the same invest-

[7] On the economics of steel mill location in South India, see W. A. Johnson, *The Steel Industry of India* (Cambridge, Mass: Harvard University Press, 1966), ch. 9.

ment.[8] Meanwhile, the allocation of foreign exchange for maintenance imports was relatively less than it would have been otherwise. Because preference was given to foreign exchange for new government investment, and new private investment was often permitted if a foreign collaborator could be found to supply the capital, large amounts of capacity remained idle because of the shortage of imported raw materials, while additional capacity continued to be built. The shortage of maintenance imports particularly affected the privately owned plants, since for bureaucratic reasons and because of a belief in the greater importance to the economy of the government plants, government plants found it easier than private ones to get foreign exchange allocated for maintenance imports.

The devaluation in 1966 had two immediate effects on those relationships. First, the rupee cost of all new investment was raised significantly. Second, the devaluation was associated with a large volume of non-project aid, which was allocated more generously than previously for maintenance imports for a large number of important industries. New government investment projects with a large foreign exchange component therefore had to be cut back, since the costs of any such projects rose sharply, and tax revenues and amounts that were budgeted for development did not increase. Meanwhile, since the private sector plants had been more heavily handicapped by the shortage of maintenance imports, they probably benefited to a larger extent than the public sector by the greater volume of non-project aid and its easier availability. For both of these reasons devaluation was sharply criticized as an anti-socialist step that was meant to benefit the private sector, and aroused strong opposition as such both in the Congress Party, where it was seen as a price-raising step before an election, and in the "left-wing" parties.

What the effect of the elections will be on the influence of socialist ideology in economic policies is uncertain. Much will depend on the choice of Prime Minister, but it will also depend on the elected membership of the Congress Party in the Lok Sabha about which nothing is known at the time of writing. The fact that the Swatantra Party will be the foremost opposition party could lead to a greater willingness to use the market and price mechanism as a basis for allocations of resources and distribution of goods. The Planning Commission has lost a great deal of prestige over the past few years, and its role is likely to decrease even further, especially since some of its leading members and staff officials have, or will be, leaving it. On the other hand, the Communist parties have increased in strength and they are more likely to raise the issue of "departing from socialism." If in fact the conservative parties do appear to be uniting into some form of opposition party the Congress Party may move further to a socialist position to oppose this new party

[8] W. A. Johnson was particularly helpful in formulating this point in this particular fashion.

and to attract supporters from the communist and socialist oppositions. Under any circumstances the new national government, with a much reduced majority, is likely to be far more sensitive than in the past to potential charges of yielding to foreign pressures or influences. Such charges may in fact have contributed to the losses and the election defeats of the Finance Minister and the Agriculture Minister, who were most closely associated with the decision on devaluation and fertilizer. In the future this heightened sensitivity may circumscribe the range of alternative economic policies a new government can consider.

Finally, what are the implications of the events of the past two years and the Indian election results for future U.S. policies toward India? It is clear that the internal Indian political situation in the next few years will be more sensitive and less stable than it was in the past. This will call for great patience on the part of the U.S. Government before seeming to exert any pressure. It is also clear that any U.S. influence on Indian policies will play only a very minor role in India's future political development. In effect this means that the United States should depend to a greater extent than it apparently has in the recent past upon technical persuasion and advice that is not associated with pressure, in seeking to influence India's economic policies. The first edition advocated a greater participation by the United States in Indian economic planning and general economic policy decision-making. It argued that for this there was an urgent need to improve the economic strength of the U.S. mission in India; that mutual discussions of Indian economic policies and programs on a technical-economic level be carried on; and that the quantity of aid at the margins should be decreased or increased, by withholding or adding aid for certain specific sectors or projects, in order to influence the allocation of investment resources and Indian sectoral economic policies, in ways considered economically desirable on clearly non-ideological grounds.

Apparently aid has been used to influence Indian economic policy. J. A. Lukas writes ". . . [Over] the last eighteen months the United States has used its food and economic aid to press the Indian government for some important policy changes. The leverage played an important role in getting India to grant concessions to foreign investors in the fertilizer industry, to take steps to increase agricultural production, to lift some strangulating controls on the economy, to liberalize imports and to devalue the rupee."[9] The ability of the United States government to use aid in this fashion was strengthened by India's great need for food as a result of the poor crops, and by the decision to stop the flow of aid of all types to both India and Pakistan because of the fighting between them in late 1965.

The use of aid to influence economic policy, however, has had costs for Indian-American relations that reduce the benefits already achieved, and

[9] *The New York Times*, January 26, 1967, p. 16.

those hoped for in the future from those policy changes. There was a great increase in Indian criticism of the United States, which must have been expected. Somewhat more significantly, perhaps, this criticism contributed to the decision of the Indian government to reconsider its previous acceptance of the setting up of an Indian-American educational foundation with PL 480 funds. Most important, the criticism that Messrs. Subramaniam and Mehta, the ministers handling food and economic policy, yielded to American pressure on agricultural policy and devaluation undermined the already weak political strength of those two ministers and Mrs. Gandhi's government. This in turn weakened the ability of those ministers in particular, and of the government in general, to implement the policies adopted or to consider new policies before the election, and may have contributed to the losses suffered by the Congress Party in the elections. Thus although the first edition argued that the United States, by its use of economic aid, could strengthen the position of various groups within the Indian government who advocated economic policy changes the United States regarded as desirable, in fact the leverage exerted by the United States may have politically weakened those groups, as well as the Congress government as a whole. In the future, because of its consequences upon a less stable Indian political system, it will be more dangerous to use or appear to use leverage with the aid program, and more desirable to rely upon persuasion and technical discussion. This will of course require maintaining the U.S. mission to India at its present high level of technical economic competence.

At the same time developments within the United States will weaken the attractiveness of aid as a source of leverage. With present congressional and public attitudes toward the foreign aid program as well as conflicting claims from Vietnam and domestic programs, it is very unlikely that U.S. economic aid to India will increase in the near future, and it may even decrease in real terms.[10] This will mean that it will be even more difficult than before to use the inducement of more aid, or easier terms, to influence India to change existing policies. This will reduce the leverage that can be exerted with the aid program, and make the Indian government even more vulnerable to the charge of yielding to pressure without receiving a satisfactory *quid pro quo*. Thus, even without the recent election results, the political costs to an Indian government of appearing to yield to foreign pressure would have risen steeply at the same time that the benefits diminish.[11]

[10] I do not think this is desirable; there are strong economic arguments both for increasing aid to India and for untying its use as much as possible. See on this, among others, I. M. D. Little and J. M. Clifford, *International Aid* (London, Allen and Unwin, 1965); E. S. Mason, "Economic Development in India and Pakistan," Harvard University Center for International Affairs, *Occasional Paper in International Affairs,* No. 13 (September, 1966).

[11] While I am writing about the United States, many of these considerations would apply to the exercise of influence by other sources of aid, such as the World Bank.

In the face of these constraints, within both India and the United States, much greater stress should be placed in U.S. economic policy toward India upon the trade element in the "trade and aid" combination, although a significant aid program will unquestionably continue. Some possibilities of using trade as a policy instrument were indicated in the first edition, and these as well as others should be explored in the United States. This exploration might be followed up by negotiations with India to reduce U.S. trade barriers to Indian goods.[12] Opening U.S. markets wider to Indian goods would force greater efficiency and lower costs upon Indian industries if they are to compete effectively in the American and other international markets, and thus help to contribute to the policy ends of the aid programs, without the need to exercise pressure. A greater dependence upon trade would also reduce the patron-client relationship, which arises from the granting of large amounts of aid, and which inevitably becomes demeaning for the recipient, thereby creating ill-feeling. Such a tariff reduction program would, however, have effects upon American industries and create political dissatisfaction in areas and industries facing greater Indian competition. It would be necessary to introduce programs to soften the problems of transition within the United States, and for this the U.S. Government could ask for concessions from India.

The difficulties of exerting U.S. influence upon the Indian government in the future make it imperative that the United States define its interests and establish its priorities with respect to its goals in India and South Asia with greater clarity than in the past. Maximizing the rate of India's economic growth is only one of those goals; others are, or could be, supporting political stability and democracy in India, protecting India against external threats from China, and discouraging further fighting between India and Pakistan. It is not certain that the United States should use whatever influence it has from its aid program for economic development alone. It is important that the use of leverage be thought through to insure that it doesn't jeopardize goals of higher or equal urgency.

Finally, the events of the past two years have made it more obvious than ever that the future of India depends in large part upon developments within India. On many of these, such as most obviously the weather, but also the communal and state developments, and the relation of the states to the Center, U.S. actions will have little or no effect in the near future. The United States must recognize its modest capabilities, even though American policies, including aid, provide a necessary support without which India's problems would be insurmountable for any friendly democratic Indian government. American policies and aid programs in India are a calculated risk on which the payoff can be very great; but they may also fail to achieve all, or even

[12] I am not discussing the role of trade policies of other developed countries, or exports of other developing countries.

many, of the desired results. The United States cannot avoid this risk, since as the world's wealthiest and most powerful democracy, it assumes responsibilities for other struggling democracies, such as India, that it cannot shrug off, both for reasons of its own national interests, and reasons of humanity.

March, 1967 G. R.

Appendixes

Appendix A

The Theoretical Framework—A Social System In Equilibrium and Disequilibrium

In this appendix I present the general framework within which the textual discussion proceeded and define my terms. This is not a general theory of the working of a society; it is meant to provide a necessary and minimum but workable set of concepts. They will inevitably be crude, they will ignore refinements and qualifications, but they will be sufficient for this analysis. This does not aim at any originality, for it is based on a wide range of reading and discussion.

The key concept is that of a social system. The term *social* implies that we are dealing with more than individuals, we are dealing with individuals in groups or groups in a larger environment. The term *system* implies that these individuals or groups interact in some meaningful and consistent fashion that can be described. In this particular analysis the major interest is in the interaction of groups, and less in the function and roles of individuals within groups, partly because the individuals derive either all or a significant part of their functions and roles from their membership in the group.[1] These groups interact in order to perform a set of functions, let us say the economic functions, within society; and in performing this general set of functions each particular group has a particular function.

In the performance of its function, each group has a role that identifies that group in relation to other groups; and within each group the members have different roles associated with their functions. In effect the roles clothe the function, identifying the group or person performing the function in relation to both other individuals within the group and other groups within the system.

Apart from the performance of functions by its members, a social system also requires a culture—a framework of ideas, possibly a religion, which serves to legitimize the role relationships among the groups, and the individuals within the groups, performing the functions. This culture also performs an educative role, since by the learning of it the members learn both their positions within society—their functions and their roles—and the legitimacy of their positions. Thus the broad culture and the nar-

[1] I shall be referring throughout only to groups, assuming that a group includes the individuals within it, but recognizing that there are differences.

rower ideology that is a part of the culture serve to rationalize the social system.

Every social system must have three major functional dimensions if it is to operate successfully. One dimension is economic, since the members of the system must be assured of some means of meeting their minimum subsistence and health requirements; some method of distribution and exchange, so that all groups performing the specialist functions that exist in any complex economy have some basis for performing their functions in relation to other groups; and, associated with this, some system of rewards that will relate what is received to function and role in the system. To the extent that the system must depend on other systems for goods or services, a method of exchange among systems must also be part of the economic dimension.

A second dimension of a social system is the political. This is the structure of government and police within the system which either ensures that the accepted interrelationship among groups is maintained, or that changes within the system in all spheres occur in a more or less legitimate and peaceful manner. The ultimate source of political power is the force that the government, formal or informal, is able to use to crush transgression. It is relatively rare that an effective government needs to use this ultimate sanction openly, but the knowledge of its existence can fulfill the same purpose. As part of the political dimension, any social system must also provide some opportunity for legitimate change within the system and some method of ensuring continuity, or it will break apart from its inadaptability to inevitable changes. A social system must also provide for dealing and conflict with other systems or other areas with similar systems, and this is in part a political function.

The third dimension of every social system is that of culture and ideology, which may be defined as a ritual dimension. This provides the method of ensuring that the traditions of the system, including its religion, are passed on to all the members of the system; ensuring that the culture adapts to environmental changes, and of controlling the adaptations so that they remain compatible with the system—that is, of distinguishing between legitimate modifications and "heresy."

In the economic sphere the determining factor in a group's economic function in a system is its "power, or lack of such, to dispose of goods or skills for the sake of income." [2] This is determined by the group's control of income-yielding assets—in an agricultural society, primarily land—but it would also include control of other types of property, or

[2] H. H. Gerth and C. Wright Mills (eds.), *From Max Weber* (New York: Oxford University Press, 1946), p. 181.

of various skills that yield income for their performance. The income derived from these properties and skills yields economic power and position. Weber, following Marx, restricts the use of the term *class* to this economic sense, and defines a class as "any group of people that is found in the same [economic power] situation." [3]

Political power on the part of a group is fundamentally dependent on the force—physical, economic, and social—that a group is able to exercise with respect to other groups, whether it is exercised directly by the group itself, by its direct influence over other groups, or by its control over the political machinery and institutions of the state, and thus indirectly and ultimately over other groups. If force is the ultimate political sanction of the system, then a group's political function is determined either by its own direct control of force, including its own contribution of pressure upon the system's political ruler, or by its influence upon the ruler of the state who controls the use of force.

Finally and not of least importance, there is the ritual position of the group. This is determined by the position of the group within the formal cultural organization of the social system, which was defined above as the organization responsible for passing on and interpreting the ideology, including the religion, of the system. I stress formal organization, since in every society much of its culture is passed on and interpreted informally, for example, within the family. However, all societies of any complexity also have a formal organization or group whose function is specifically to make the definitive interpretation of the culture, to teach the population—especially the youth—the characteristics of the culture, to decide what is "heresy," and to recommend sanctions against heretical behavior. It is the function of the political power within the system to carry out those recommendations, but the political power may hand over this duty to the religious organization. Within these formal organizations there is a hierarchy of position, and any one group's position depends on the finality of its interpretation of the system's *scriptures*. (In this presentation I have overemphasized the rationality of this system; there are inevitably many contradictions and illegitimacies that are tolerated within the ritual organization.)

To understand the functioning of such a system it is useful to start by positing a static system—that is, one not influenced by the passage of time. I will first define a static system in stable equilibrium.[4] By *equilibrium* I mean that there is a direct relationship between a group's rank in one sphere, reflecting its power in that sphere, and its rank and power in other

[3] *Ibid.*

[4] I am deliberately using the concept of "equilibrium," because of its convenience, although I admit the argument over its value, and the difficulty of definition.

spheres. Thus it is located in the entire system, although the group's major function within the system will probably be in only one sphere.[5] *Stability* implies that relatively small movements away from a group's interdependent ranking in each of the three spheres will set up a response within the system that will reestablish an appropriate relationship. I shall present a relatively simple example of this type of equilibrium centering about a powerful ruler. Any small step to weaken his position, let us say to deface his statue, in turn requires only a minor police action, which is however sufficient to re-enforce his rule by showing his strength. This example makes it possible to see another type of equilibrium, an *unstable equilibrium,* in which an initially slight movement away from the original equilibrium does not lead to a return to the original equilibrium, but rather leads to a still further movement from that relationship. An example of such a change would be an attack upon the ruler from another group, which induces many other groups to abandon him, so that he is replaced by another ruler. This may not further upset the stability of the system as a whole, if the successor is produced within the same system that produced the original ruler, and if the successor accepts the original system.[6]

In contrast to this concept of a static system in equilibrium, independent of changes over time, I now introduce the element of time—that is, a *dynamic system.* It is possible to postulate a series of discrete, relatively small changes in each of the spheres; occurring slowly and spread sufficiently over time, the entire system is able to establish a new equilibrium position after each change, and the changes in each sphere can be treated in isolation from changes in the rest of the system. In a stable equilibrium a change in any one sphere gives rise to appropriate changes in other spheres, which lead over time to either the re-establishment of the original

[5] At the top of the system there are likely to be most important exceptions to this: a priest-king may have great powers in all spheres.

[6] For further explanation of the concept of equilibrium several economic analogies would be useful. In economics a stable market equilibrium is defined as such a relationship of given demand and supply schedules (curves on a diagram), that slight movements away from the intersection of the schedules (or curves) brings into play a reactive mechanism that causes a return to the intersection. Myrdal, in his work on economic change, has stressed the unstable character of some types of equilibria: when an original worsening (or improvement) of a relationship leads to a cumulative deterioration (or betterment) of a group's position vis-à-vis that of other groups. These concepts of equilibrium are theoretical, they are not historical; and an equilibrium position is not, necessarily or usually, one that is observed. Furthermore it is frequently impossible to predict in fact, as distinguished from theory, if the original position is stable or unstable—the results of the disturbance determine the type of stability. Finally, I wish to stress that the concept of equilibrium as used here is meant to have no normative connotations whatsoever.

equilibrium or a movement to a new equilibrium position.[7] In an unstable equilibrium the effect upon the system of such a change in one sphere is to establish either an improved or worsened position for the affected group in its own sphere, depending on the direction of instability. This in turn gives rise to further changes which accelerate the movement from equilibrium in this and in other spheres. Since I am now postulating a slow rate of change, the ascending or descending spiral is halted at the next stage and the old system continues, but modified. For such a controlled change to occur successfully certain assumptions must be made: on one hand, with respect to the character of the single change, that it is relatively isolated, that it accepts the system as a whole, and that it occurs at a slow enough speed so that the system can in fact absorb it; and on the other hand, with respect to the flexibility of the system, that the relationships within the system are not so brittle that any push away from equilibrium leads to its collapse, either from the ascent or the descent of any one component. The concept of flexibility also implies that groups within the system are peacefully willing to accept relatively slight gains or losses in their position in one sphere, or they are willing to wait and work through other dimensions to reverse the loss either within the one sphere or the system as a whole. The role of intellectuals and religious leaders is highly significant in dealing with these shifts and in the willingness of a group to accept or reject them, and thus they are important in determining the flexibility of the system.

The final type of movement in this classification is the drastic change or changes, completely upsetting the equilibrium system. This may be a drastic change in any sphere of a character that neither recognizes nor accepts the relationships in the existing system. Frequently accompanying the initial drastic change there may be a rapid series of changes in all spheres, each of which calls for adjustment by the existing system, but which together occur in such number and at such a speed that the existing system does not have time to adjust to all the changes.[8]

The English conquest of India is an example of drastic, contradictory change. In effect it substituted, on the political level, a different system for the one then functioning. The new system was entirely different in its structure of group relationships, in its values and ideology, and in the

---

[7] This might be compared to the "comparative static" analysis of economic theory; for example, where shifts of the entire demand and supply schedules over two periods lead to new equilibrium positions at the end of the shift.

[8] To use another analogy from economic theory, this type of change might be considered as the equivalent of a structural change as distinguished from the marginal type of change described earlier.

institutions it introduced. However, after the major change of rule, although the British introduced many additional changes they introduced them at a pace that permitted the existing system to operate by either adjusting to them or quarantining them; in effect the two systems operated side by side. The Hindu system, defined as the system found by the British in India, was in an unstable equilibrium—changing, giving ground, on the defensive against British ideas and institutions. Consequently new group relationships had been established, but it had far from collapsed as of Independence. A more rapid type of change is exemplified by the changes that have accompanied and followed Independence in India—legal, political, social, economic, and religious changes—all occurring within fifteen years, striking at the old system with a wide variety of pressures, and in the process going far to create a new society containing parts of the old and the new systems interwoven.

Before I go on to the Indian case specifically it is necessary to set forth a set of definitions for *gain* and *loss*. By gain or loss I mean a change in the position of a group, or in the rewards accruing to a group, as a result of a change in the economic, social, or ritual dimensions of a social system. In economics the term is reasonably precise and measurable to the extent that it is measured by monetary units: it is an increase in the real money return to a group or an individual performing an economic function. (Allowing for price changes over the period raises problems.) To the extent that there are nonmonetary gains in the form of increased leisure or nonincome perquisites, measurement of even economic gain becomes very difficult. Furthermore, I have spoken so far of absolute money gain, which may be measurable. The measurement of relative shifts in economic well-being becomes even more difficult since it requires comparing the gains of one group with those of another; this in turn raises problems of defining the group, and brings in dimensions of *felt* position that the economist normally does not handle. It is obvious that absolute gain is no necessary guide to the direction of relative gain, since they may move in opposite directions; it is also obvious that although a group may on an average gain in either absolute or relative terms, many individuals within that group may lose.[9]

[9] It is at times claimed that the inhabitants in underdeveloped countries are less interested in material gain than those in developed countries. The reasons given for this may be custom or religion, and this statement is often made with certain moral overtones. In my own experience with Indian businessmen, I have found little evidence for this claim; and the work done by anthropologists and economists in rural areas among Indian peasants also gives little support for the argument. The most readily available discussion of this point, with reference to India, is Milton Singer, "Cultural Values in India's Economic Development," *The Annals of the American Academy of Political and Social Sciences*, CCCV (May,

In the social and political spheres the measurement of gain or loss by a group is far more difficult. Great changes are noticeable. It is obvious that if a ruler is overthrown and replaced, the old ruler has lost politically and the new one has gained. But political and social gains or losses are not measurable in any units such as money income, and the shifts in position are all relative; gains to one or more groups involve losses to other groups. It is possible that the losses in, say, the economic sphere may be compensated for by either gains in, say, the political sphere or, more likely, by maintenance of position in other spheres of activity. A group may be willing to accept the shift in its position among these spheres of activity depending upon the character of the shift, and also upon which of its functions and roles it feels is most important. One of the main determinants of this is the culture and ideology of the group, and of the system. If, for example, the ideology of an aristocratic group lays most stress upon political and social responsibility and power, then that group may be willing to accept economic losses to retain them: this may be what occurred in England or Japan. But, if discrepancies in position among the various spheres becomes too great; if at the same time each group develops an ideology supporting its claims for greater position in—let us say—the political or social spheres to go with its increasing power in the economic sphere; if there is little or no opportunity to bridge the gap between a group's present position and its desires or expectations as expressed in a general ideology or by significant groups of intellectuals—the results may be violent revolution.

One of the clearest descriptions of such a disequilibrium position is contained in the opening pages of Georges Lefebvre's classic, *The Coming of the French Revolution.*

At the end of the Eighteenth Century the social structure of France was aristocratic. It showed the traces of having originated at a time when land was almost the only form of wealth, and when the possessors of land were the masters of those who needed it to work and to live. It is true that in the

1956), pp. 81–91; and the discussion of the same theme by John Goheen, M. N. Srinivas, D. G. Karve, and Milton Singer, "India's Cultural Values and Economic Development: A Discussion," *Economic Development and Cultural Change,* VII, No. 1 (October, 1958), pp. 1–12. With respect to India I think it can be assumed correctly that economic gain or loss does act as an incentive for personal and group action, and that the desire for economic improvement is a goal on the part of both the individual and the state. In my opinion the economic motivation may in fact be stronger in underdeveloped than in affluent societies, precisely because the former societies are nearer to the margin between bare subsistence and destitution, and the inhabitants can therefore less afford to neglect economic factors. At the same time this economic motivation probably operates within a very different social and institutional context than in the developed countries, or even in the developed areas of an underdeveloped country.

course of age-old struggles . . . the king had been able gradually to deprive the lords of their political power and subject nobles and clergy to his authority. But he had left them the first place in the social hierarchy.

Meanwhile the growth of commerce and industry had created, step by step, a new form of wealth, mobile or commercial wealth, and a new class . . . the bourgeoisie. . . . This class had grown much stronger . . . because it proved highly useful to the monarchical state in supplying it with money and competent officials. . . . It had developed a new ideology. . . . The role of the nobility had correspondingly declined; and the clergy, as the ideal which it proclaimed lost prestige, found its authority growing weaker. These groups preserved the highest rank in the legal [and social] structure of the country, but in reality economic power, personal abilities and confidence in the future had passed largely to the bourgeoisie. . . . The Revolution of 1789 restored the harmony between fact and law.[10]

And it is for this reason that De Tocqueville concluded that, "it is not always by going from bad to worse that a society falls into revolution. . . . The social order destroyed by a revolution is almost always better than that which immediately preceded it, and experience shows that the most dangerous moment for a bad government is generally that in which it sets about reform." [11]

In India the caste system is a social system in the sense used in this appendix. Caste, which traditionally structures a society on the basis of the birth of its members, determines a group's ritual function in religious observances and its interrelation with other groups in ritual and social behavior; it determines a group's economic function, its economic relation with other groups in the geographic area, and its rewards; it determines in large part a group's political function and role in an area. Finally the caste system was at first and still is a very important part of the Hindu religion, which supports and is supported by it, and there is some question whether the two are separable.

In theory, caste and class position would be inseparable in the caste system in equilibrium. With the erosion of the ideological and material bases of the caste system, mostly under British rule, a disequilibrium arose between caste and class position, and simultaneously the caste system faced strong intellectual attack. In rural India, however, at Independence the caste system was still ruling, although class elements were both operable and noticeable within rural society; in urban India the class elements had probably become overriding, modified by caste and ethnic factors.

---

[10] Georges Lefebvre, *The Coming of the French Revolution* (Princeton: Princeton University Press, 1947), pp. 1–2.

[11] De Tocqueville, *L'Ancien Régime*, p. 186.

Appendix B

Bibliography on Caste

The caste system in India is an extremely complex system, and there is a very extensive literature on it. In Chapter 3 I presented certain broad generalizations based in part on the general books on caste in India. An even more useful source are the numerous village and regional studies and a few urban studies carried on by anthropologists in many areas of India. From the village studies it is possible to conclude that, although the regional differences in the functioning of the rural caste are many and great, they represent variations on a major underlying theme. The purpose of this appendix is to present some of the textually uncited published sources from which I reached the generalized conclusions.

Part I indicates some general bibliography; Parts II and III list articles and books on caste in rural and in urban India, respectively. In this appendix I list books and articles that describe the role of the caste system either in its more traditional form or as it functioned in 1950. I make no claim either for the inclusiveness or definitiveness of the following list.

I. CASTE: SOME GENERAL LITERATURE

About the general functioning of caste—apart from the usual sources, such as the classic studies and interpretations by G. S. Ghurye, J. H. Hutton, Max Weber, and Abbé Dubois—I have found the following more recent works of great value:

Beteille, A., Y. B. Damle, S. Shahani, and M. N. Srinivas. *Caste: A Trend Report and Bibliography.* Oxford: Basil Blackwell, International Sociological Association, n.d.

Bose, N. K. *Peasant Life in India: A Study in Indian Unity and Diversity.* (Anthropological Survey of India, Memoir No. 8 [1961].)

Damle, Y. B. *Caste: A Review of the Literature on Caste.* M.I.T. Center for International Studies, 1961. A general annotated bibliography and review.

Karve, I. *Hindu Society—An Interpretation.* Poona: Deccan College, 1961. An analysis of the origins and position of caste.

Kolenda, P. M. "Toward a Model of the Hindu Jajmani System," *Human Organization* (Spring, 1963). The entire issue of this magazine is devoted to the Jajmani system.

Leach, E. R. "Introduction: What Should We Mean by Caste," *Aspects of Caste in South India,* ed. E. R. Leach. Cambridge: Cambridge University Press, 1960.

Mandelbaum, D. G. "Concepts and Methods in the Study of Caste," *The Economic Weekly Annual,* XI (January, 1959).

Marriott, M. *Caste Ranking and Community Structure in Five Regions of India and Pakistan.* (Monograph Series No. 23.) Poona: Deccan College, 1960.

Neale, W. C. "Reciprocity and Redistribution in the Indian Village," *Trade and Market in the Early Empires,* ed. K. Polanyi, *et al.* Glencoe, Illinois: The Free Press, 1957.

Singer, M. (ed.), *Traditional India: Structure and Change.* Philadelphia: The American Folklore Society, 1959. See especially the articles by D. Ingalls, H. Lamb, J. Hitchcock, and W. N. Brown.

Srinivas, M. N. "The Dominant Caste in Rampura," *American Anthropologist,* LXI (February, 1959), 1–16.

———. *Religion and Society Among the Coorgs of South India.* Oxford: Clarendon Press, 1952. Srinivas' book and article contain definitions of the concepts of "sanskritization" and "dominance."

II. CASTE IN RURAL INDIA: SOME AREA ANALYSES

The following items describe caste in a given region or reach general conclusions from specific area studies.

Bailey, F. G. *Caste and the Economic Frontier.* Manchester: Manchester University Press, 1957.

Beidelman, T. O. *A Comparative Analysis of the Jajmani System.* Locust Valley, New York: J. J. Augustin Incorporated Publisher, 1958.

Berreman, G. D. "Caste and Economy in the Himalayas," *Economic Development and Cultural Change,* X (July, 1962). This is an analysis of the economic aspects of caste in a Himalayan village.

Bose, N. K. (ed.). *Data on Caste in Orissa.* (Anthropological Survey of India, Memoir No. 7.) Calcutta: 1960. This presents a very detailed description of the functioning of various castes in Orissa.

Carstairs, G. M. *The Twice Born.* London: The Hogarth Press, 1957. This is an analysis by a psychologist of high caste peasants in a village of Rajasthan.

Dube, S. C. *Indian Village.* Ithaca: Cornell University Press, 1955. This is very able study of a central Indian village.

Gough, K. "Caste in a Tanjore Village," in Leach (ed.), *op. cit.* This is a perceptive look at caste in a South Indian community.

———. "Review of a Comparative Analysis of the Jajmani System," by

T. O. Beidelman, *Economic Development and Cultural Change,* IX (October, 1960). An excellent criticism by a leading anthropologist with great experience of India.

GOI Planning Commission, Programme Evaluation Organisation. *Leadership and Group in a South Indian Village.* New Delhi: The Manager of Publications, 1955.

Indian Society of Agricultural Economics, The. *Bhadkad.* Bombay: 1957. A comparative study of a village of Gujerat in 1915 and 1955. There are only a few such studies.

Karve, D. D. *The New Brahman.* Berkeley and Los Angeles: The University of California Press, 1963. This presents the changing picture of the Brahmins in Maharashtra.

Kumar, D. "Caste and Landlessness in South India," *Comparative Studies in Society and History,* IV (April, 1962), 337–363. This article traces the relationships among landlessness, low caste, and slavery in South India from the mid-nineteenth century to the present.

Lewis, O. *Village Life in North India.* Urbana: University of Illinois Press, 1958. This book emphasizes the role of factions in village life in North India.

Majumdar, D. N. *Caste and Communication in an Indian Village.* Bombay: Asia Publishing House, 1958. This contains a study of a North Indian village by one of India's foremost anthropologists.

Mayer, A. C. *Caste and Kinship in Central India.* Berkeley and Los Angeles: University of California Press, 1960.

———. "The Dominant Caste in a Region of Central India," *Southwestern Journal of Anthropology,* XIV, No. 4 (1958), 407–427. This book and the preceding article contain analyses of caste in general, and more specifically in a Central Indian village.

Mukherjee, R. *The Dynamics of a Rural Society.* Berlin: Akademie-Verlag, 1957.

Sarma, J. "The Secular Status of Caste," *Eastern Anthropologist,* XII (Dec., 1958–Feb., 1959), pp. 87–106.

Srinivas, M. N. (ed.). *India's Villages.* West Bengal Government Press, 1962. Originally articles in the *Economic Weekly* (Bombay) from 1951 to 1954.

Wiser, William H., and Charlotte V. *The Hindu Jajmani System.* Lucknow: Lucknow Publishing House, 1958.

III. CASTE IN URBAN INDIA

There is far less published literature on India's urban social structure than on its rural society. Much of it is contemporary and describes the present situation. The following items are some of the readily available studies of urban social structure that contain a historical perspective, especially of the period around Independence.

GENERAL STUDIES

Crane, R. I. "Urbanism in India," *American Journal of Sociology,* LX (July, 1954—May, 1955), 463 ff.

Pocock, D. F. "Sociologies: Urban and Rural," *Contributions to Indian Sociology,* No. 4 (1960), 63–81. This and the preceding are two brief and general review articles on India's urban history. They will be useful for anyone desiring a brief background to the subject.

Punekar, S. D. "The Middle Class Today," *The Economic Weekly,* XI (February 28, 1959), 317–318.

Srinivas, M. N. "The Indian Road to Equality," *The Economic Weekly,* XII (June, 1960), 867–872.

Turner, Roy, (ed.). *India's Urban Future.* Berkeley and Los Angeles: University of California Press, 1962. The article by R. Lambert is particularly useful for a general picture of the sociology of Indian cities today. The book itself is indispensable for understanding Indian urban life.

SPECIFIC STUDIES OF SOCIAL STRUCTURE IN DIFFERENT CITIES OR URBAN AREAS

Bose, N. K. "Some Aspects of Caste in Bengal," *Traditional India: Structure and Change,* ed. M. Singer. Philadelphia: The American Folklore Society, 1959. This presents data on caste and occupation for a small Bengal town and for all Bengal.

Chauhan, D. S. "Caste and Occupation in Agra City," *The Economic Weekly,* XII (July 23, 1960), 1147.

Driver, E. D. "Caste and Occupational Structure in Central India," *Social Forces,* XLI (October, 1962), 26–31. A study of caste and occupation in Nagpur in Central India.

Gist, N. "Caste Differentials in South India," *American Sociological Review,* XIX (April, 1954), 126. A study of caste and occupation in urban Mysore in South India.

Hart, H. "Urban Politics in Bombay," *The Economic Weekly,* XII (June, 1960), 983 ff.

James, R. "Labor Mobility, Unemployment and Economic Change," *Journal of Political Economy,* LXVII (December, 1959), 545–559.

Lambert, R. D. *Workers, Factories, and Social Change in India.* Princeton: Princeton University Press, 1963. A study of the characteristics of the factory labor force in Poona. This and the Niehoff study below are the only two such studies I know.

Morris, M. D. "Recruitment of an Industrial Labor Force in India, with British and American Comparisons," *Comparative Studies in Society and History,* II (1959–1960), 305–328.

Niehoff, A. *Factory Workers in India* ("Publications in Anthropology," No. 5). Milwaukee: Milwaukee Public Museum, 1959. This is one of the few book-length studies of the sociology of factory labor in India. It presents a most valuable picture of factory labor in Kanpur.

Sen, S. N. *The City of Calcutta.* Calcutta: Brookland Private Limited, 1961. One of the studies sponsored by the Planning Commission on major Indian cities.

STUDIES OF RURAL MIGRATION TO URBAN AREAS

Epstein, T. S. "Industrial Employment for Landless Labourers Only," *The Economic Weekly,* XI (July, 1959).

Gould, H. A. "Sanskritization and Modernization," *The Economic Weekly,* XIII (June 24, 1961), 945–950; XIV (January 13, 1962), 48–51.

Appendix C

Numbers in Middle Class Groups, 1961
(Table 21)

EMPLOYERS

The National Sample Survey shows a rise in the proportion of employers to total urban population from 0.28 per cent in 1952 to 0.46 per cent in 1957/1958. Such a rise would not be surprising as a result of the economic development that did occur. Using the higher proportion and applying it to the figure of 79 million, the total urban population in 1961, yields a figure for the number of employers of about 350,000.[1]

WHOLESALERS AND RETAILERS, n.e.c.

The National Sample Survey estimates that in 1957/1958, 12 per cent of the urban gainfully employed population were engaged in wholesale and retail trades. If this ratio is applied to the estimated 26 million urban employed in 1961, the total number in retail and wholesale trade would equal 3.1 million. In 1959 there were approximately 120,000 commercial establishments, restaurants, theaters, and so on, covered under various state registration legislation. I think it reasonable to estimate that most of these were probably located in urban areas; I also assume that about 100,000 of these were owned by employers, who are already included in the previous group, and should therefore be subtracted from the 3.1 million to reach a total of 3.0 million.[2]

PROFESSIONAL PERSONNEL

The Third Plan estimated that the number of doctors increased in proportion to the population from 1951 to 1961; applying this proportion to the number of professional medical and health personnel would result in an increase of about 30 per cent for urban India or by about 70,000. The total number of teachers in all India rose from 750,000 to 1,369,000

[1] GOI Cabinet Secretariat, *Tables with Notes on Employment and Unemployment in Urban Areas*, "National Sample Survey," No. 63 (Calcutta: 1962), Table 7, p. 36; later references are to Table 43, p. 87.

[2] NCAER, *Urban Income and Saving*, p. 29; GOI Ministry of Labour and Employment, *Indian Labour Statistics* (Simla: 1961), p. 46.

during the same period. Applying the percentage increase for the total number of teachers in India to the base figures for urban teachers in the early 1950's results in an estimate of about 700,000 urban teachers in 1961. There is no estimate of changing numbers of journalists or newspapers. In 1951, I estimated about 100,000 journalists on the basis of a crude assumption of ten journalists per newspaper. On the basis of an almost 50 per cent increase in the circulation of English and Hindu papers alone from 1952 to 1957, I estimate a proportionate increase in the number of journalists.

It is probable that engineers would be classified among administrators, executives, and technicians rather than professionals, since almost all are employed by government or private firms. I have no independent estimate of the number of lawyers who would not be included under government employees. However, in 1960 it is probable that there were less than 100,000 lawyers in India altogether, based upon the number of law students and graduates in 1939/1940 and 1949/1950. I assume that there were about 50,000 lawyers not elsewhere classified in 1950/1951 and that the increase was in proportion to the population growth—probably not higher, given the limited opportunities and stress upon other types of employment.[3]

ADMINISTRATORS, EXECUTIVES, TECHNICIANS (INCLUDING CLERKS)

In 1957/1958 this group included approximately 12 per cent of the urban employed labor force. Applying this ratio to the 26 million urban employed in 1961 results in a total size of this group of 3.1 million. A major employer of this group of personnel is the Indian government or the state governments. Total government employment rose from 3.5 million to 6.6 million from 1951 to 1960, and of this, employment in urban areas rose from about 2.5 million to 4.7 million. I estimated in Chapter 2 that there were 300,000 office and administrative workers employed by the central government in 1951, and Healey has estimated that this figure rose to above 400,000 in 1960. State and local government employees are not broken down by types; but if I assume the same ratio of urban white-collar state and local government employees (excluding teachers) in 1960 as in 1951, the total of such employment in 1960 would be 1.4 million. Any change in these totals from 1960 to 1961 would be minor. Therefore the total urban middle class government employment in 1961 equalled 1.8 million employees.

[3] *The Third Plan,* pp. 575 and 662; *Report of the Working Journalists Committee,* pp. 11–15; B. B. Misra, *The Indian Middle Classes,* p. 333.

In both 1951 and 1960 the number of clerks was 70–75 per cent of the total number of central government employees. I have applied the 70 per cent ratio also to the state and local government employees, and to the figure for the banking and insurance employment, which I believe is closer in character to government employment than to manufacturing.

With respect to this group in manufacturing, the Planning Commission estimated that 8 per cent of the 3.2 million workers in registered factories in 1956 were classified as professional, technical, administrative, executive, clerical, and sales workers. I assume that this same proportion held for the 2.9 million workers in registered factories in 1951 and the 3.8 million in 1961. I also assume that approximately 70 per cent of these white-collar employees are in urban areas, reflecting the national urban to rural proportions for employment in manufacturing. In 1956 clerks were 55 per cent of the total employees in this class in manufacturing, and I also assume this held in 1951 and 1961. Applying these proportions to the totals, I get the figures for employment in this class, both total and the number of clerks, which are presented in Table 21.[4]

BANKING AND INSURANCE EMPLOYEES

The Bank Disputes Tribunal of 1962 estimated total employment in banks in 1959 at 70,000; and it was determined that employment in insurance companies in 1961 was 58,000. I assume that there was a small increase in bank employment to 75,000 by 1961, and that in 1961, 120,000 employees out of the total in both activities were urban white-collar workers. Both banks and insurance companies are largely urban phenomena, although during the development effort some attempt has been made to extend these institutions to rural areas.[5] The size of the clerical staff was estimated by applying the 70 per cent ratio in government employment.

THE EDUCATED UNEMPLOYED

This group is not in the middle class as an economic income class, since its members receive no income. However, many members of the group, because of family background, education, and expectations, consider themselves as members of the middle class. As such, many would prob-

[4] GOI Planning Commission, *Occupational Pattern in Manufacturing Industries,* p. 11, Table T.5.1; *Indian Labour Journal,* XIV, No. 1 (January, 1963), p. 82, Table 1. "Registered Factories" are those employing either more than 10 workers and using power, or more than 20 workers and not using power.

[5] On this see The Bank Disputes Tribunal, *Award on the Industrial Disputes;* and *The Monthly Abstract of Statistics,* "Insurance," March, 1963, pp. v and vi.

ably be unwilling to accept employment unless it were in a white-collar, middle-class job, and this group should therefore be included within the middle class.

The number of educated unemployed registered with employment exchanges rose from 216,000 in 1955 to approximately 600,000 at the end of 1961, and to over 700,000 by September, 1962. It is estimated that those registered at exchanges are 40 per cent of the total unemployed in this group, which would yield a total of 550,000 in 1955 and of 1.0 million in 1961. Of this total, based on the proportions registered in the employment exchanges, approximately 10 per cent were college graduates, 12 per cent had passed their intermediate examinations (comparable to junior college graduates), and the remainder were matriculates (comparable to high school graduates). It is of some interest that in a survey of the employment experience of Delhi University graduates of the 1950 and 1954 classes, approximately 20–25 per cent of the job-seeking alumni suffered unemployment for more than one year during the subsequent five or six years, and 15–20 per cent of the graduates seeking work were unemployed for a year or more before getting their first position.[6]

[6] On the educated unemployed see GOI Planning Commission, *Outline Report of the Study Group in Educated Unemployed* (Delhi: January, 1956), chap. 3; GOI Ministry of Labour and Employment, *Employment Survey of the Alumni of Delhi University* (New Delhi: 1962), pp. 114–123; *Monthly Abstract of Statistics,* January, 1963.

Appendix D

Numbers in Working Class Groups, 1961 (Table 24)

Table 24 presented a summary of the numbers in various groups of the urban working class, exclusive of agricultural employees, in 1961. The explanation of the 1961 figures is as follows.

OPERATIVES AND ARTISANS (INCLUDING SUPERVISORS) AND HOUSEHOLD INDUSTRY

In Chapter 2 I estimated the total number of workers in these groups at 5,540,000 in the early 1950's, divided into 3.5 million in manufacturing and 2.0 million in household industry on the basis of National Sample Survey data. This is probably within the margin of error of the 1961 Census total estimate for 1951 of 5.8 million workers in both urban manufacturing and household industry, which unfortunately is not broken down, and it would be of little value to try to reconcile the figures into a more exact agreement.

The Census indicates that the urban labor force in manufacturing alone rose to 5.5 million in 1961, or a rise of 45 per cent compared with the estimated Census base for 1951. If one estimates a similar proportionate rise in the number of operatives and artisans (including supervisors) in manufacturing, this raises the early 1950's estimate of 3,500,000 to approximately 5,100,000 in 1961. Of these, approximately 2,700,000 would be in urban registered factories.[1]

MANUFACTURERS OF FOOD PRODUCTS AND TEXTILES

This group includes largely household workers and many self-employed. The 1961 Census estimates 2.1 million engaged in urban household industries in 1961.

SERVICE WORKERS

I estimate the increase in such workers would be in the same proportion as the increase in "other service" workers in urban areas from 1951 to

[1] For the method of computation of those in registered factories in urban areas see Appendix C—Administrators, Executives, Technicians.

1961, or by 10 per cent. The employment of this group would therefore equal 3.4 million in 1961.

HAWKERS AND THE LIKE

The estimate for the 1950's was based on the National Sample Survey figure of 2.4 per cent of the urban labor force in this group. The National Council of Applied Economic Research (NCAER) estimated that 4.1 per cent of the urban households in the cities they surveyed in 1960 were engaged in work of this nature.[2] On the assumption that the proportion of hawkers, peddlers, and so on would be lower in smaller towns, I estimate a proportion of 3.5 per cent of the urban households in these occupations, and a similar proportion of the urban labor force of 26 million in 1961. This proportion would yield a total figure of 900,000 hawkers and peddlers.

BUILDING INDUSTRY WORKERS

In the early 1950's the number of workers engaged in the building industry was approximately 50 per cent of the total number of urban construction workers as indicated by the Census. Applying the same proportion to the 1961 Census yields a total of 500,000 workers in the building industry, not elsewhere classified.

UNSKILLED LABOR

The NCAER study indicates a rise in the proportion of unskilled laborers to the total labor force—this study is based on urban households—from 8.1 per cent in the early 1950's to 10.6 per cent in 1960.[3] This proportion is unlikely to differ greatly in towns above and below 10,000. Applying the 10.6 per cent ratio to the 1961 urban labor force yields a total of 2.8 million unskilled laborers.

UNEMPLOYED (OTHER THAN EDUCATED UNEMPLOYED)

It is extremely difficult to measure unemployment accurately in India; there are serious problems of definition, of disguised unemployment, of registration, of movement from rural to urban areas, and so on. It is far beyond my purpose to discuss this measurement problem, and I therefore

[2] NCAER, *Urban Income and Saving*, p. 49. This survey was limited to cities with populations of 10,000 or more, covering about 75 per cent of the total urban population.

[3] *Ibid.*

simply assume that the number of applicants on the live employment registers provides some measure of urban unemployment (or rather those looking for jobs), on the ground that only in urban areas is there likely to be significant use of these registers.

The total number of applicants on the live registers rose from 330,000 at the end of 1951 to about 700,000 at the end of 1955, to 1.8 million in December, 1961. If from these 1955 and 1961 totals the known figures of 215,000 educated unemployed on the rolls in 1955 and 600,000 on the rolls in 1961 are subtracted, the result is an increase in the number of uneducated unemployed from almost 500,000 in 1955 to 1.2 million in 1961, or a 140 per cent increase.[4] There is unquestionably a large number of unemployed who do not register; in 1956 this number of unregistered unemployed was placed at 75 per cent of the total unemployed, compared with 60 per cent of the educated unemployed[5] but this proportion has probably decreased since 1956. This underregistering means that the 1.2 million figure for this group of unemployed workers is probably a minimum estimate and the total of urban unemployed was probably closer to 2.0 million than to 1.0 million in 1961.

---

[4] *Monthly Abstract of Statistics,* January, 1963, pp. 2 and 69.
[5] *The Indian Second Five-Year Plan,* chap. v.

Appendix E

State Incomes from 1951–1961

As I mentioned on p. 195, the only measure of communal economic gains or losses, apart from the study by R. K. Hazari, is the movement of state incomes over the decade. Since the states reflect the linguistic divisions within India, strongly since 1956 and somewhat less so before then, they also reflect major communal groups.

Unfortunately, although many of the states have made estimates of their income, nearly every state has a long lag in presenting the data. Furthermore, there are incomparabilities in definitions of sectors and methods of computing sectoral incomes; and there are differences between the national totals and the sum of the state totals. It is probable, too, that the problems previously noted with respect to gaps and weaknesses in national income data are at least as serious in the states as they are in the central government, in part because the average level of competence in the Central Statistical Office is likely to be somewhat higher than in the corresponding state offices, given relative salary and prestige levels.[1] With this warning I am going to make use of the data that exist, since they probably indicate reasonably well for my purpose the relative position of the states with respect to their economies, their trends in incomes, and their development since Independence. The various linguistic groups within India, more or less represented by the states, have gained by development; I shall try to indicate to what extent these changes can be explained, even crudely, by such factors as the comparative experience of peasants working land (as postulated by Kusum Nair); comparative size of holdings (as postulated by K. N. Raj); and relative levels of state government development expenditures. Appendix Tables E-1 and E-2 summarize a series of relevant data with respect to development in the Indian states during the decade. The tables should be used cautiously, given all the qualifications of the data, as a measure of trends, certainly not to distinguish fine differences among the states. Table E-3 presents various ranking of the states by different criteria of development.

[1] See the references to the quality of Indian economic statistics in Chapter 7; B. F. Hoselitz and M. D. Chaudhry, "General Survey of State Income Studies in India," prepared for the Fourth Indian Conference on Research in National Income (December, 1962) provides a detailed analysis of state income estimates for 1955/1956. The statistics for many of the states have had to be adjusted for boundary changes occurring after 1956.

APPENDIX TABLE E-1

STATE POPULATION 1950 TO 1960 AND ESTIMATED PLANNED EXPENDITURE ON DEVELOPMENT, SECOND PLAN

State	Population (millions)				Estimated Expenditure in Second Plan by State (millions of rupees)	
	1950	1951	1959	1960	Total	Per Capita 1961
Andhra	30.8	31.1	35.9	36.0	1,806	50
Assam	8.9	8.8	10.5	11.9	632	53
Bihar	38.3	38.8	43.6	46.5	1,769	38
Bombay	47.3	48.3	56.9	60.2		60
Maharashtra		32.0		39.6	2,140	54
Gujerat		16.3		20.6	1,468	71
Kerala	13.2	13.6	16.4	16.9	790	47
Madhya Pradesh	25.8	26.1	29.0	32.4	1,455	45
Madras	29.5	30.1	34.6	33.7	1,862	55
Mysore	19.0	19.4	23.0	23.6	1,387	59
Orissa	14.5	14.7	16.1	17.6	894	51
Punjab	15.8	16.1	19.0	20.3	1,514	75
Rajasthan	15.7	16.0	18.8	20.2	999	49
Uttar Pradesh	62.3	63.2	71.7	73.8	2,283	31
West Bengal	26.0	26.3	28.9	34.9	1,558	45
India [a]	356.3	361.1	415.0	439.2		

Notes:

[a] India totals do not represent totals of the states. Jammu and Kashmir data are not available for the earlier year; nor is there much data for union territories and other areas.

SOURCES:

Population: *Census of India, 1961,* for 1951 and 1961; K. N. Raj, *The Economic Weekly,* February 4, 1961, p. 260, Table 16, for 1950 and 1959 estimates.

Incomes—total and state: *Ibid.,* p. 296, Table 16.

APPENDIX TABLE E-2

TRENDS IN SOME ASPECTS OF STATE DEVELOPMENT DURING THE PLAN DECADE

State	Income					
	Total (millions of 1948/1949 rupees)		Percentage Change	Per Capita (1948/1949 rupees)		Percentage Change
	1949/1950	1958/1958		1949/1950	1958/1959	
Andhra	7,035	9,362	33.1	229	261	14.1
Assam	3,046	3,406	11.8	343	324	−5.8
Bihar	7,646	10,239	33.9	200	235	17.7
Bombay	12,894	17,804	38.1	273	313	14.9
Maharashtra						
Gujerat						
Kerala	3,097	3,952	27.6	234	242	3.2
Madhya Pradesh	6,593	9,240	40.1	256	319	24.6
Madras	6,744	9,324	38.3	229	269	17.6
Mysore	3,531	4,721	33.7	186	205	10.1
Orissa	2,730	3,199	17.2	188	199	5.9
Punjab	5,225	7,713	47.6	331	406	22.9
Rajasthan	2,701	4,484	66.0	173	239	38.2
Uttar Pradesh	16,321	20,603	32.8	262	287	9.7
West Bengal	9,190	11,501	25.1	353	398	12.8
India	88,400	117,000	32.4	248	282	13.6

APPENDIX TABLE E-2 (Con't.)

TRENDS IN SOME ASPECTS OF STATE DEVELOPMENT DURING THE PLAN DECADE

State	Agriculture				Distribution of Workers in Non-primary Employment			
	Indexes of Farm Output (base period = 100)			Annual Percentage Change Total ÷ Number of Years	1951		1961	
	Base Period	1950/51	1960/61		Million	Per cent	Million	Per cent
Andhra	1949/50	105	123	1.7	3.3	28.8	5.2	28.4
Assam	1956/57	93	96	0.3	0.6	15.0	1.1	21.7
Bihar	1949/50	72	104 [a]	4.9	1.8	13.9	3.8	19.8
Bombay					6.7	31.5		
Maharashtra	1956/57	81 [b]	120(P)	5.3			5.2	27.9
Gujerat	1949/50	111	190 [c]	8.9			2.6	30.8
Kerala	1952/53	107 [d]	120 [d]	2.4	1.9	43.9	3.0	53.1
Madhya Pradesh	1952/53	110 [e]	147(P)	4.8	2.3	18.6	3.0	17.8
Madras	1947–1950	100	137 [f]	4.6	3.2	36.3	5.7	36.7
Mysore	1952/53	129 [g]	141	1.3	1.9	28.3	2.8	26.4
Orissa	1949/50–1958/59 [h]			1.1	1.5	27.1	1.8	24.4
Punjab	1956/57	101 [i]	120	6.3	2.0	32.7	2.4	35.1
Rajasthan	1952–1956	84 [j]	127(P)	6.1	2.0	24.2	2.0	20.4
Uttar Pradesh	1948/49	108	142	3.2	6.3	23.6	7.0	24.2
West Bengal	1949/50–1958/59 [h]			1.1	4.0	43.3	4.8	41.2
India	1949/50	96	139	4.5	38.0	27.3	52.2	27.8

Notes:

(P) Indicates provisional or preliminary.

[a] 1959/1960. [b] 1951/1952. [c] 1958/1959. [d] 1953/1954 and 1958/1959.

[e] 1953/1954. [f] 1958/1959. [g] 1953/1954.

[h] Income from major crops for Orissa and West Bengal rose by 11.0 per cent and 11.2 per cent, respectively. See: K. N. Raj, "Some Features of Economic Growth," p. 259, Table 7.

[i] 1957/1958. [j] 1952/1953.

SOURCE:

"Worker Distribution," *Census of India,* 1961, Appendix I, Tables 11–13, pp. 402–413.

APPENDIX TABLE E-3

RANKING OF VARIOUS STATES IN TERMS OF DEVELOPMENT MEASURES[a]

State	Income			Rates of Growth		Farm Output Annual Rate of Change	Structural Change in Employment	Plan Expenditure	
	Total 1949/50 1958/59	Per Capita 1949/50	Per Capita 1958/59	Total Income	Per Capita Income			Total	Per Capita
Uttar Pradesh	1	5	6	9[b]	10	7	5	2	13
Bombay	2	4	5[b]	5	6	1	—	1	2
West Bengal	3	1	2	10	8	11	—	6	10
Bihar	4	10	11	6	5	4	2	5	12
Andhra	5	8	8	8	7[b]	9	—	4	7
Madras	6	8	7	4	4	6[b]	6	3	4
Madhya Pradesh	7	6[b]	4	3	2	5	—	8	10
Punjab	8	3	1	2	3	2	4	7	1
Mysore	9	12	12	7	9	10	—	9	3
Kerala	11	7	9	11	12	8	3	12	9
Assam	12	2	3	13	13	13	1	13	5
Orissa	13	11	13	12	11	11	—	11	6
Rajasthan	10	13	10	1	1	3	—	10	8

Notes:

— Indicates negligible.

[a] I have deliberately not computed rank correlations of these variables since the basic data are so weak, and such correlations would give a spurious air of accuracy. However, the table does show shifts in relative position and general relationships that will be useful to analyze.

[b] The indicated states are just above the national average for that variable, where such an average makes sense. Thus the states with lower numbers than the indicated ones would be even more above average; and vice versa.

SOURCE:

Farm Output Indexes: Economic Survey of Indian Agriculture, 1960–61, Table 2.1, pp. 78–89.

The Punjab and Rajasthan are the two states that have shown the most rapid rate of economic growth. The states with the most rapid increase in farm output are Gujerat in the former Bombay state, Punjab, Rajasthan and Maharashtra; and with farm income such a large proportion of the total, it is not surprising that these states show a tendency toward fastest growth rates. There is no obvious relationship between shifting employment and rising income, which is not surprising considering the relative constancy of the structure of employment on a national level. Per capita Plan expenditure by the states seems to have only a slight relationship to rates of growth in per capita income, but since these expenditure figures do not include either private investment or central government expenditures in the states—these data are either unavailable or unpublished—the relation of state public expenditures to growth would not be direct.

The high rate of increase of farm output noted in the states provides some support for Kusum Nair's hypothesis, since the peasants in states that have shown most rapid growth tend to be from those peasant castes or groups that have a tradition of working the land themselves. K. N. Raj also points out the tendency for larger farms to be operated in Northwest and West India, and if larger farmers have been able to adopt newer techniques and expand farm output more rapidly than smaller ones, Table E-3 would tend to support his argument on the relationship between size of landholding and rate of increase in farm output. There are, however, other specific factors that unquestionably influence the variable rates of growth of specific states; for example, the concentration of the hard-working Sikh and Punjab refugee communities in the Punjab has unquestionably contributed to the growth of the Punjab. It is generally agreed that the states of Madras and of old Bombay, and its new components, are among the best governed in India in terms of speed and efficiency, and this has proved attractive to private investment. The existence of active entrepreneurial groups—within Bombay City in Maharashtra, Ahmedabad in Gujerat, Calcutta in West Bengal, Madras and Coimbatore in Madras, and Rajasthan with its position as the home of the Marwari business community and its nearness to New Delhi—have contributed to the growth of those states; and in the larger cities there are important economies of scale in location that have attracted industry and private investment.

In general terms it appears that the development that has occurred has not had much effect on the ranking of the states in terms of either total or per capita incomes, but it would seem to indicate a trend toward more rapid growth rates in the northwest than elsewhere. Apart from this tendency, growth does not seem to be much more rapid in the north than in the south, the main line of linguistic division within India, which casts doubt on the southern claim of discrimination; but what may be most important, the great state of Uttar Pradesh, by far the largest in India, the center of the Hindu language, and a center of the past power of the Congress, ranks relatively low in all the growth indexes. Similarly, West Bengal and Kerala are relatively slow in growing. The relationship between this varying pattern of growth rates and such questions as political stability within the state Congress parties, the role of opposition parties, and the level of government efficiency are indirect; certainly they are not straightforward in any causal connection. I believe that in West Bengal and Kerala these economic and political problems are closely related to the changing positions of the lower middle class and in Uttar Pradesh to the increasing tension among caste groups. However, the political form in which these effects show themselves vary: the Congress party in West Bengal is one of the least faction-ridden in India; in Kerala and Uttar Pradesh the state Congress parties are very unstable but they are also unstable in the fast growing states of the Punjab and Rajasthan.[2] The economic factors are subsumed under a wide variety of other forces when directly explaining Congress party functioning in the individual states, and the political forces outlined in Part II of the text are probably better guides than economic factors to understanding the party's operations in those states.

---

[2] Far more work is necessary on state politics than has been done up to now to reach any firm conclusions on the relation between economic and political change in the states.

Index

Index

Selected RAND Books

BAUM, WARREN C., *The French Economy and the State,* Princeton, N.J., Princeton University Press, 1958.

BERGSON, A., *The Real National Income of Soviet Russia Since 1928,* Cambridge, Harvard University Press, 1961.

BRODIE, BERNARD, *Strategy in the Missile Age,* Princeton, N.J., Princeton University Press, 1959.

CHAPMAN, JANET G., *Real Wages in Soviet Russia Since 1928,* Cambridge, Harvard University Press, 1963.

DINERSTEIN, HERBERT S., *War and the Soviet Union: Nuclear Weapons and the Revolution in Soviet Military and Political Thinking,* New York, Praeger, 1959.

DOLE, STEPHEN, and ISAAC ASIMOV, *Planets for Man,* New York, Random House, 1964.

DORFMAN, ROBERT, PAUL A. SAMUELSON, ROBERT M. SOLOW, *Linear Programming and Economic Analysis,* New York, McGraw-Hill Book Company, Inc., 1958.

HALPERN, MANFRED, *The Politics of Social Change in the Middle East and North Africa,* Princeton, N.J., Princeton University Press, 1963.

HIRSHLEIFER, JACK, JAMES C. DEHAVEN, and JEROME W. MILLIMAN, *Water Supply: Economics, Technology, and Policy,* Chicago, The University of Chicago Press, 1960.

HITCH, CHARLES J., and ROLAND MCKEAN, *The Economics of Defense in the Nuclear Age,* Cambridge, Harvard University Press, 1960.

HSIEH, ALICE L., *Communist China's Strategy in the Nuclear Era,* Englewood Cliffs, N.J., Prentice-Hall, Inc., 1962.

JOHNSON, JOHN J. (ed.), *The Role of the Military in Underdeveloped Countries,* Princeton, N.J., Princeton University Press, 1962.

JOHNSTONE, WILLIAM C., *Burma's Foreign Policy: A Study in Neutralism,* Cambridge, Harvard University Press, 1963.

LIU, TA-CHUNG, and KUNG-CHIA YEH, *The Economy of the Chinese Mainland: National Income and Economic Development, 1933–1959,* Princeton, N.J., Princeton University Press, 1965.

LUBELL, HAROLD, *Middle East Oil Crises and Western Europe's Energy Supplies,* Baltimore, The Johns Hopkins Press, 1963.

MCKEAN, ROLAND N., *Efficiency in Government through Systems Analysis: With Emphasis on Water Resource Development,* New York, John Wiley & Sons, Inc., 1958.

MEYER, J. R., J. F. KAIN, and M. WOHL, *The Urban Transportation Problem,* Cambridge, Harvard University Press, 1965.

MOORSTEEN, RICHARD, *Prices and Production of Machinery in the Soviet Union, 1928–1958,* Cambridge, Harvard University Press, 1962.

NOVICK, DAVID (ed.), *Program Budgeting: Program Analysis and the Federal Budget,* Cambridge, Harvard University Press, 1965.

PINCUS, JOHN A., *Economic Aid and International Cost Sharing,* Baltimore, Md., The Johns Hopkins Press, 1965.

QUADE, E. S. (ed.), *Analysis for Military Decisions,* Chicago, Rand McNally & Company; Amsterdam, North-Holland Publishing Company, 1964.

RUSH, MYRON, *Political Succession in the USSR,* New York, Columbia University Press, 1965.

SPEIER, HANS, *Divided Berlin: The Anatomy of Soviet Political Blackmail,* New York, Praeger, 1961.

TRAGER, FRANK N. (ed.), *Marxism in Southeast Asia: A Study of Four Countries,* Stanford, Calif., Stanford University Press, 1959.

WHITING, ALLEN S., *China Crosses the Yalu: The Decision To Enter the Korean War,* New York, The Macmillan Company, 1960.

WILLIAMS, J. D., *The Compleat Strategyst: Being a Primer on the Theory of Games of Strategy,* New York, McGraw-Hill Book Company, Inc., 1954.

WOLF, CHARLES, JR., *Foreign Aid: Theory and Practice in Southern Asia,* Princeton, N.J., Princeton University Press, 1960.

WOLFE, THOMAS, *Soviet Strategy at the Crossroads,* Cambridge, Harvard University Press, 1964.

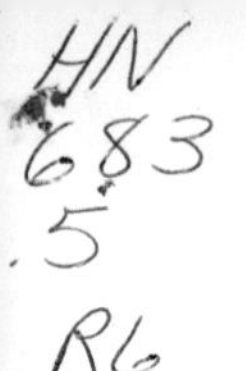

Rosen, George

Democracy and Economic
Change in India

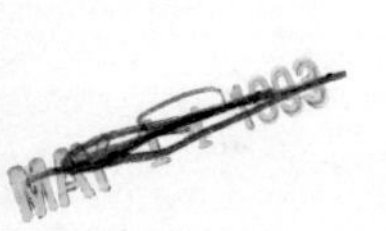